Conestoga Highway

By Geoff Miller

ISBN 978-0-9971860-3-1

Acknowledgments: I'd like to thank the numerous family members and friends who read drafts of the manuscript and provided helpful comments. In particular I want to thank Doug Critchfield, good friend and talented surveyor, for providing technical help related to his profession and for guidance on New Mexican culinary issues.

Disclaimers: This is a work of fiction. Any similarities to real people are entirely coincidental. In addition, I've taken considerable editorial license in my descriptions of the respective County offices in Taos and Mora, New Mexico. Places and protocol have been fabricated to fit the plot. For this I beg the forgiveness of the citizens of these charming towns.

For: Erica, Nicolas and Pamela

I

It was dark that night. Clouds swept past the crescent moon periodically allowing just enough light to poke through for non-nocturnal creatures, such as *Homo sapiens*, to perform the most rudimentary of tasks. But that was all he was doing. With bloodied hands he hastily, violently, dug and scratched at the hard soil, soil so densely packed that progress was frustratingly slow. He soon found a sharp, flattened rock, something he could plunge and drag through the shallow hole, and the pace of his work improved. After more than three hours of backbreaking labor the job was almost done—this most precious of tasks. Tears streaked through the dust that had been fixed to his skin by perspiration. The night was cold but even in his loose cotton shirt he was sweating profusely.

Lifting the body of his dead wife, he laid her as gently as he could into the shallow grave, kissing her one last time before drawing the makeshift shroud over her face. Before he began to fill in the hole, he pulled a small cedar box from his poke and placed it on her breast, between her folded hands. It was finely constructed with silver hinges and clasp, and inside it contained the small gold crucifix that had been his gift to her on their wedding day.

"En el nombre del padre, del hijo, y del spiritu santo," he muttered as he made the sign of the cross and gazed skyward, his face contorted in anguish. With a heavy heart he completed the job and began to roll and pile the rocks back into place. He then rebuilt the cairn that supported the large wooden cross that the priest had set there almost five years earlier during a time of great joy and promise. Although the native people would certainly remove the cross, he was fairly certain they wouldn't realize what lay underneath the rock pile. The soil from near the bottom of the grave was moist, but by morning any exposed, loose dirt resulting from his work would have dried and desiccated to match the appearance of the surrounding surface material. The rocks were too large for the beasts of the forest to move, and he fervently hoped nothing would desecrate the site until, perhaps, he or one of his sons could return and give her a proper Christian burial.

Javier Bustamonte stood and said one last prayer before shouldering his pack and stumbling towards the relative safety of the nearby forest. A glance to the northeast revealed dark shapes passing back and forth in front of the flames now engulfing the structure that, until today, had been his residence, the hacienda he'd built and improved with his own hands over the last seven years. For now all he could do was to rejoin the others in their flight to safety. They were headed southward to the port city where he'd left his sons and where soldiers of the king would repulse any further aggression from the locals who were attempting to drive the Spanish from their homeland. The year was 1680 and the violence, in what would one day become the northern part of the state of New Mexico, would be referred to in history books as the "Pueblo Revolt."

II

Five men on horseback cantered toward the dry creek bed, the concerned looks on their faces indicating the importance of their endeavor. The lead rider reined in his mount and signaled the others to do likewise. He tipped his hat back, leaned on his saddle horn, and swept the horizon, taking in the rugged southern New Mexican landscape dotted with sagebrush, rabbit bush, and the occasional twisted juniper tree. Two buzzards drifted through the cloudless blue sky, riding thermals created by uneven heating of rocks that rimmed the distant hills. The man dismounted and walked bowlegged to the crest of a low escarpment. His gait was brisk, conveying urgency, his spurs jangling in concert with the crunch of worn pointed boots scuffling across the hard-packed desert soil. He knelt briefly and examined the ground before turning to address his compatriots.

"You was right, Luke. Looks like they come through here sometime this mornin', 'bout a hour ago's my guess, two men on horseback and maybe five, six head of cattle."

The man named Luke slowly rode his horse to the front of the party. Tall and handsome, he had a firm, chiseled chin, high, aristocratic cheekbones, and piercing blue eyes. His demeanor and the looks he received from his colleagues left no doubt as to who was the leader of this group. Sitting tall in the saddle he paused to contemplate the new information he'd been given. When he spoke, his voice was booming, authoritative, "Wes, you and Sammy track to the south along the arroyo. Bart, you, Jimmy and I'll head west over the hills. These rustlers'll be movin' slow as long as they're drivin' stock. If we ride all night we can set us up a little surprise party for 'em in manana!"

One of the other men on horseback began to speak, but was interrupted by a bald-headed man wearing dark glasses, seated just off camera. "Hold it. Cut. Dirk, you stay put; the rest of you can take five." He turned towards his assistant, a slender woman with short, dark hair. His tone was exasperated. "Shit, nobody read through the script with him this morning did they?"

"No, Mr. Bivens, I'm sorry, there just wasn't time. Besides, you know how defensive he gets when he thinks we're trying to coach him."

Bivens nodded in resignation. It wasn't Wendy's fault that Dirk Stetson, the star of his latest movie, was such a fuck-up. "Hey, Dirk, let's do a close-up while you say that last line again. By the way, buddy, nothing in the line rhymes with 'banana.' The word in question is *mañana*, pronounced 'ma-nya-na.'"

Stetson shook his head and frowned, "I don't know, Andre, Luke's supposed to be a rancher, a cow-puncher; I think it's out of character for him to speak in Mexican."

"I don't have time to argue with you, Dirk. Just say the freakin' line like it's written, okay?" Abruptly he began barking orders to his production staff: the camera, sound, makeup and set crewmen who would attempt to turn this short but pivotal scene into box office gold.

Andre Bivens felt he had a great script for his new movie 'Conestoga Highway.' This project provided a golden opportunity to recover from the long seven-year drought since his one true success, the breakout film that had secured his reputation as a respected Hollywood director. What a wild ride he'd been on back then. That movie, *his* movie, 'Autumn Leaves,' had been the critic's choice for best film of the year. It was a dark, somber, psychological thriller set in a quiet New England town, a solid film that had also finished a close runner up in balloting for the best picture Academy Award. Although it had not attracted many moviegoers during its initial limited release, it had immediately garnered high marks for crisp direction; suspenseful, provocative editing; and great sound work. The good reviews drove people into theaters when the movie went global, and it ended up making a tidy profit for its investors.

Unfortunately, the two major projects he'd worked on since then had lost money. Reviews had been favorable, and Andre's reputation as a director had remained solid, but audiences had been drawn away to more popular films of a different genre. According to live action purists like Andre, it seemed that anything involving sophisticated special effects or computer-generated graphics was virtually guaranteed to be a hit: dialog, plot and social relevance notwithstanding. The studios where Bivens had strong contacts had been either slow or reluctant to acquire the necessary resources to produce such films, and they, along with anyone connected with them, were hurting for business.

Just over a year ago, a little-known novelist and screen writer handed Andre the script for 'Conestoga Highway.' Why he hadn't immediately committed it to the circular file was, in itself, unusual, but something on the first page, the setting under the very first slugline, had held Bivens's attention long enough to pique his continued interest. He'd kept reading the script straight through to the end, finishing it in the early hours of the morning, all the while growing ever more excited about its potential. On his recommendation, producers, studio executives, and, most importantly, investors, had bought into the idea of making the movie.

As a child Andre had spent most of his free time in a new, and by mid-twentieth century standards, 'ultra-modern,' air-conditioned Cineplex in his hometown shopping mall. He would devour popcorn and jujubes while sitting transfixed before the myriad of offerings Hollywood sent his way. From the age of fifteen, he was certain what he wanted to do with his life—work in the movie industry—a dream he'd now lived, with passion and without regret, for the past forty years. And what had captured his soul and started him on that journey was the epic "western." As a young boy he'd sat spellbound as Gary Cooper and Burt Lancaster rode the plains hunting bank robbers, fighting claim jumpers, and rescuing damsels in distress from runaway stage coaches. The fists of John Wayne defended the rights of honest cattle ranchers, and the blazing guns of Clint Eastwood saved entire towns from evil gangs of extortionist gunslingers and rogue rustlers. While children nowadays escape into virtual worlds created by computer programmers, back in the late '50s and early '60s young boys like Andre Bivens stared up at the silver screen and imagined themselves as heroes of the old west, fighting alongside movie stars who were bigger than life—stars like Dirk Stetson.

Stetson was once considered the heir apparent to John Wayne, but the sudden demise if the "western" had derailed his path to becoming a Hollywood leading man. One minute he was at the top of his profession and the next he was forced to accept lesser roles, mostly as a typecast character actor. When things really slowed down, the occasional commercial stint would help pay the bills; he had briefly played a young Marlboro Man until cigarette ads were banned. Dirk was tall and ruggedly handsome, but, unfortunately, he was also dumb as a post. Unaware of the man's diminished intellectual status, Andre had lobbied to sign him into the lead role as "Luke" in this revivalist western, but he soon regretted that decision as Stetson invariably screwed-up scene after scene, stumbling over the simplest of lines. The filming schedule had slipped badly due to Dirk's mistakes. In a valiant attempt to keep the film moving and within budget, the screenwriters had masterfully cut Dirk's dialog to little more than two short sentences per scene, and a personal assistant, a voluptuous, eager, wide-eyed blonde named LaVerne—herself no rocket scientist—had been hired to read through each scene with Dirk a minimum of four times before it shot. Alas, the plan had been a failure. LaVerne was spending considerable time in Dirk's plush, air-conditioned trailer, but it soon became evident that very little rehearsal was taking place behind closed doors. Bivens knew that the situation threatened his authority as director, but there wasn't much he could do about it. Production was so far along now that he was stuck with the stupid, temperamental actor until the end of the project.

Andre stood and began walking to the editing trailer. A glance back at the set buoyed his spirits somewhat; the scene was already underway again. Wendy was one of the few employees on the crew who could deal with Dirk's insecurity, and whatever performance she could get out of him could conceivably be whipped into shape by his talented post-production staff. He entered the trailer to speak briefly with the young man sitting in front of the editing console. A blast of cool air hit him as he stepped into the room and closed the door. His editor, Chris, sported a set of sophisticated-looking earphones and was staring intently at four flickering monitors. The limited counter space was littered with empty diet Dr. Pepper cans and overfilled ashtrays. The place reeked of stale cigarette smoke.

"Shit, Andre," the young man said without looking up from his work, "splicing that last scene together is gonna be awkward. We may have to put something else in there to break it up—a background shot or something." He paused as he slowed the replay to a few frames per second, "and maybe Bruce can help us out a bit with some sort of sound change."

"Well, see what you can do," Andre said with a sigh. Chris and Bruce were among the best editing and sound men in the business. He knew he was damned lucky to have them as part of his crew.

"That's six beers you now owe me, Andre, and if I had any kind of social life I'd be getting pissed with all the extra hours I've been putting in."

"Glad you're not pissed just yet," Bivens said with a wink before heading back to the set—a set that was still set up and preparing for yet another take. *"Good Lord, what now?"* he wondered. As he strode back to the arroyo, Wendy, clipboard in hand, welcomed him with a smile.

"We're almost done. He's gotten the pronunciation right, but he thinks mañana is the name of the place where the posse's going to rendezvous. I set him straight. This should be the last take."

Without a word Bivens plopped into his director's chair and lit a cigarette. He took several deep drags, sat back and closed his eyes, trying to relax as he listened to Wendy's attempt to finish this nightmare of a scene.

"No, no, Dirk, tip your hat with your 'other' left hand. Remember," she added sweetly, "the camera's off to the right, shooting you in profile. Just think of all those admiring women who want to see your handsome face."

Dirk was an idiot, but he sure looked the part, Andre thought optimistically as the cut was finally recorded for posterity. Without missing a beat the crew began to break down the equipment, hurriedly and professionally preparing for the journey to the next location on the schedule. They would do three scenes there, at a carefully selected site just outside of Abiquiu, New Mexico, the landscape made famous in the works of the artist Georgia O'Keefe. Andre was looking forward to working in that locale. It was tantalizingly close to Dirk's ranch up near Taos, which also meant he would be that much closer to Bella.

Dirk Stetson and Bella Star were husband and wife. Although it was his fourth marriage and her third, their eight year relationship was often cited in the tabloids as one of the few stable, committed marriages in all of Hollywood. Andre knew better. The marriage was really one of convenience, and both partners were old hands when it came to dealing with the paparazzi. Being married made them appear boring and much less interesting to the tabloid reporters who crawled around 'tinsel town' at all hours of the day hawking gossip and snapping impromptu photos of celebrities for their rabid fans. The Stetsons were publicly affectionate and very discrete in their affairs, and although rumors of infidelity would surface occasionally, they would be expertly squelched when the two showed up arm-in-arm at the next awards program or star-studded gala, kissing, laughing, and mugging for the cameras. The truth, Andre knew, was that aside from those carefully planned public outings they almost never saw one another.

He'd tried to get the producers to cast Bella in the role of Luke's wife. It would have generated great publicity to have one of Hollywood's most famous couples star side-by-side. The proposal, however, had been nixed immediately. Bella was beautiful and a superb actress, but enough bloom had left the rose to deny her any consideration for the part. Although Dirk was several years her senior he was male, and that of course meant age wasn't an issue when it came to casting his role. The double standard disgusted

Andre, but he knew the producers were right. Filmmaking was a cutthroat business and enough money had been invested in this work that they had to pull out all the stops. There would be a nude scene involving rancher Luke's wife shot in silhouette from far enough away so as to keep the film's PG-13 rating. In order to ensure good theater attendance among thirteen to thirty-year-old males, the woman in question had to be portrayed by an actress chosen from a select group of current heartthrobs. Unfortunately, Bella was no longer on that list.

Andre had great respect for Bella both as a person and as an actress. Years earlier they had been lovers, but their extracurricular activities hadn't affected their work. As a director he greatly appreciated Bella's complete commitment to her role whenever she was engaged in a project. For a while he'd been disappointed in her flippant attitude towards their relationship; he'd always considered it to be special until he learned that she'd bedded many of the cast and crew as well. He quickly realized it was just her nature; she was impulsive, affectionate, and blessed with unusually high libido, a true nymphomaniac. If the schedule allowed, maybe he could drive to Taos for a day and go see her, let Wendy deal with her intellectually-impaired husband while he and Bella brought one another up-to-date with news of their lives.

III

It was an impressive gate, one of the largest Denny Carswell had ever seen. Groups of rough-hewn logs on either side supported iron doors nearly fifteen feet high that towered above his truck. Capping this magnificent entryway was a metal sign that arced above the gates, sporting a plasma-cut scene of elk grazing among tall pines and with the name "Stetson" artfully engraved in the center.

Denny emerged from his truck and walked the few steps to an intercom box mounted off to the side of the gate. He punched a button labeled "call" and moments later a metallic voice emanated from the unit.

"Stetson ranch."

"Good morning, my name's Carswell. I'm here regarding a property survey."

A pause ensued, then, "Yes, Mr. Carswell, I'm Larry. We've been expecting you."

Denny heard a loud click, and the doors of the gate silently swung open. He returned to his vehicle and followed a long winding road through peaceful pine forest to Dirk Stetson's ranch house.

"English Victorian log mansion" is the phrase that sprang to mind as Denny rounded the circular gravel drive in front of the main residence. He slowed to a stop and contemplated the queer-looking building. The walls around the ground floor were constructed of large pine logs. Old and weather-beaten, they were probably remnants of the original house. The upper floor was quite different. The regularity of the siding suggested that it was stick-built with an aesthetic log veneer that someone thought would blend with the first floor. It didn't. Numerous gabled rooms poked out below the roof line, each neatly trimmed with a small porch and railing, a gaudy mix of architectural styles that seemed out of place in a rural New Mexico setting. As he stared at the building, a slender blonde-haired man stepped lightly down the front porch steps and strode with purpose to meet him. His dress was as foreign to the landscape as was the ranch house. He wore a button-down shirt, open at the collar, a light-colored dinner jacket and soft leather shoes. As he approached, Denny made out a round face sporting glasses with black plastic frames that bestowed a 'Buddy Holley, geek 1960s' look.

"Mr. Carswell, thank you for coming." The man held out a hand in greeting.

"Thank you, Mr . . . ."

"Larry Cooper, please call me Larry. I'm Mr. Stetson's publicist. I also handle most of his personal affairs including operation of this ranch. Speaking of which, we plan to sell a portion of it, and that's why we called you. Please, let's go inside and I'll fill you in on the details."

Abruptly he turned on his heels and walked back into the house expecting Denny to follow. Larry was obviously a no-nonsense type; Denny had to rush to keep up the pace as they climbed the porch steps. Larry opened one of the heavy, ornately-carved front doors and they stepped into a cool, high-ceilinged foyer. The inside of the house was nothing like the awkward façade that had greeted Denny moments earlier. As he scanned the great room that lay open in front of him he was impressed by the style and skill with which it had been decorated. Larry sensed Denny's awe and beamed with pride. He suppressed his nervous energy and paused to let his guest take in his surroundings.

Dark, weathered pine wainscoting gave way to light, polished aspen paneling that extended all the way to the top of what had to be twenty foot high ceilings. Two large chandeliers made of intertwined elk antlers hung down from above, and Denny could imagine how their glow would reflect off the walls on cold winter evenings. To either side of where he stood, winding staircases led to the upper floors, and in the back corner off to his right stood a great hearth faced by comfortable but rustic chairs set on thick carpet. The floor was heavy solid pine and the walls were artfully adorned with paintings of western landscapes and scenes of Dirk from his films, most of them showing the actor on horseback.

"Welcome to our little cabin in the woods," Larry said in understatement. He motioned for Denny to follow him towards the back of the house. They passed through another set of double doors, through a dining room containing a huge half-sawn log table, then out through sliding glass doors and onto a large outdoor patio. Larry beckoned Denny to sit at a small table shaded by a rainbow-colored umbrella. The transition from southwestern ranch house to a scene reminiscent of a Caribbean beach club was abrupt and unexpected. Numerous tables, like the one where they sat, rimmed a large kidney-shaped swimming pool. The porcelain tile surrounding the pool contained designs of shells, palm trees and flamingoes.

"I know what you must be thinking," Larry said. He swept his hand across the view of the patio. "This is Bella's domain. She just loves the beach. We have an unwritten understanding: I design what's in the house, but the pool area is strictly hers. Let me go and get a copy of the last survey of the property," he continued once his guest was settled. "It won't take but a minute. Would you care for a beverage while we discuss business?"

A glass of water would be nice, if it's not too much trouble."

"It's no trouble at all," Larry replied with a smile. He was back moments later carrying a tray containing what appeared to be a large Bloody Mary and a tall glass of ice water topped with a twist of lime artistically skewered on a swizzle stick. He had a folder tucked under one arm that he placed between them and from which he extracted a large folded plat of the estate.

"Here's what we were given the last time the property was surveyed. I wasn't yet an employee here when it was done." He shrugged, "From my layman's point of view it looks to be in order. Oh," he continued as he opened the folder again, "the survey that accompanied the deed when Dirk bought the ranch seventeen years ago is here too." He placed a three-page Xerox copy in front of Denny and settled into his chair, savoring a sip of his cocktail. "It looks like the second survey is just copied from the first. I don't know why Dirk even bothered getting a second one done. The descriptions and measurements are the same."

Denny examined the documents side-by-side. The larger document, the more recent survey, was neatly prepared using CAD software and consisted of a plat (a scaled drawing of the property) together with metes and bounds verbiage describing the boundary in prose:

> "Beginning at a stone monument in the Rio Casteneda, due south from the church in the town of Arenas Rojas, hence to the northeast on a bearing of N 38° 29' E for a distance of 2450.78 feet to an iron pin . . . to a stone pile at the top of Cerro Medio, hence along a fence bearing S 78° 10' W for a distance of 2850.00 feet to an iron pin in the Rio Casteneda, hence North and East along the thread of the Rio for a distance of 3970 feet, more or less, to the point of beginning."

The document bore the seal and license number of the surveyor, Joshua Marquez, and was dated June 3, 2009. Denny was impressed with the clarity and neatness of both the plat and the description. He had to agree with Larry; it looked like a straightforward and complete outline of the parcel.

"The Indian Pueblo over in the hills to the north, Santa Teresa, has been asking to buy the part of the ranch that bounds the arroyo, the so-called Rio Casteneda in the deed description, for several years now. Word has it they want to put a casino on the land because it would have easy access from the highway. Their latest offer was $7500 an acre." He finished his drink and set the glass down with a satisfied grunt. Denny could tell the vodka was beginning to take effect; Larry was starting to act even more outgoing and enthusiastic than before.

"Now, I'm obviously no surveyor," he continued, "but I know a good land deal when I see one. The injuns probably figure that casino's gonna be a gold mine-and-a-half to offer that kind of dough. By my calculation they'll be paying nearly a million smackers for that land. Hell, that's more'n what Dirk paid for this entire spread back in the '90s. Your job, Mr. Carswell, should you decide to accept it"—his tone imitated the opening sequence from 'Mission Impossible'—"will be to parcel off and mark the corners of the property we wish to sell. I've arranged a meeting between us, Conrad—our lawyer—and representatives of the tribe, for Thursday evening. I trust that four days is enough time to get your work done." This last statement was uttered as a statement rather than a question.

Denny was only vaguely aware of what Larry was saying. Although he was trying hard to listen to his host, at the same time he was perusing the older document he'd been given, the Xerox copy of the original deed. In other words he was multitasking, but doing a poor job of it. The deed was taking precedence. It consisted of three pages, the first of which contained typical "legalese" jargon coupled with information about where and when the deed was recorded—information he'd verify later that day or the next. The second and third pages of the deed were what held his attention. Halfway down page two began a metes and bounds description credited to a surveyor named Antonio Roybal. There was no accompanying plat. What Roybal wrote seemed to be, verbatim, what was later written on the Marquez document except for the wording of the final course that appeared on page three:

". . . from paraje San Cristobal, on the summit of Cerro Medio to the Rio Casteneda."

The difference was slight but unmistakable. Why did the last part of Marquez's description deviate from the older text when the rest was identical? The term "paraje San Cristobal" also struck a chord. An idea was beginning to gel when his train of thought was interrupted by Larry.

His host was waving his hand, and his voice was slurred, "Helloooo, Mr. Carswell, do you think you can handle the job?"

"What? Oh, I'm sorry. The deed is, um, interesting. I guess there shouldn't be a problem with subdivision of the parcel, but one never knows. Let's hope there are no surprises . . . ."

"Ah, Bella, darling. There you are." Larry rose to greet a woman who had just entered the pool area from the dining room. He gave her a peck on the cheek. "You look ravishing, as usual."

"Thank you Larry. You always know how to greet a lady. My goodness, I'd have dressed for the occasion if I'd known you had company." Denny stood and turned to face her. Her voice deepened, adopting a sultry quality, "My, you're a tall, handsome drink o' water. What's your name, sugar?"

"Denny Carswell, ma'am. I'm a surveyor. Larry here has tasked me with parting off the portion of the ranch you're fixin' to sell. You must be Mrs. Stetson. It's a pleasure to meet you." Denny sheepishly held out his hand while averting his gaze slightly. The woman who stood before him was clad in nothing but high heels, a bikini bottom, large sunglasses and a wide-brimmed sun hat. In her left hand she carried a large beach towel.

"That's me, in the flesh so to speak. I'm Bella Starr, also known as Bella Stetson to my adoring public." She giggled, removed her sunglasses and shook hands. "I'm sorry if my nudity embarrasses you, but I just can't stand tan lines—they make me look like a zebra."

Bella's nonchalant attitude about her state of undress made Denny relax somewhat. She had struck a pose that seemed to ask, "Take a good look and tell me what you think." He coughed politely not knowing quite how to act or what to say.

Weeks earlier, Pamela, Denny's wife, had gotten all goggle-eyed when she'd learned he'd been offered a job at the Stetson estate. She'd remarked that the whole world—except, of course, for Denny Carswell—knew that Dirk and Bella Stetson were one of Hollywood's leading couples. Pam's subsequent exuberant demeanor showed a side of her that he'd never seen in the forty-plus years they'd been together. An animated litany had sprung from her lips: a recitation of an impressive list of films in which one or both had played starring roles. Details of their tumultuous but committed marriage then followed as well as gossip about the movie currently being filmed in New Mexico, the one starring none other than Dirk Stetson. Denny never would have guessed that his wife paid so much attention to people in the entertainment industry. Nonetheless, he was now grateful for one bit of information she had provided to him. Bella had apparently starred in the movie 'Twice

Beguiled,' a rather simple but entertaining movie he and Pam had seen nearly twenty years earlier. He vaguely remembered the film's very beautiful female protagonist.

"Well, I must say that you look every bit as lovely in person as you did in 'Twice Beguiled.' It was a very enjoyable film." The truth of the matter was that he didn't immediately notice any resemblance between the woman standing in front of him and the actress in the film. However, his reference to her beauty was emphatically genuine. In a word, she was gorgeous. He knew she had to be in her late fifties from what Pam had said, but she'd been able to cheat father time amazingly well. High heels made her appear taller than she was; Denny guessed that without them she'd stand perhaps five-seven or five-eight. She had full lips and a slightly upturned mouth which gave her a perky, fresh, girl-next-door look. When he studied her eyes a vestige of memory from the film flashed through his mind. They were a dazzling sky-blue, topped by light brown, perfectly trimmed eyebrows and surrounded by smooth skin devoid of crow's feet or any telltale signs of aging. She also kept herself in good shape. The muscles in her legs and arms were firm, and, despite a small mid-rift paunch, she had a physique many twenty-year-olds would die for. Although she was beautiful, there was also an unnatural look about her, one indicative of extensive plastic surgery. She'd had a facelift or two, and her lips, well, Denny had read somewhere that there was a procedure for injecting them with something to make them fuller; Mother Nature had clearly received some help in that department. It was her breasts, though, that figuratively jumped up and cried 'surgical enhancement.' Bella seemed to be thrusting them at him, obviously proud of their size, shape and firmness.

"Why, thank you, Denny," she replied with pleasure. "I'm glad you enjoyed it. That film was a lot of fun to make. And please call me Bella, 'Mrs. Stetson' makes me feel old." She winked at him and returned her sunglasses to her face. "Now, if you'll excuse me, I have to work on my tan. The October sun can be so fickle; one has to grab it when one has the chance." She brushed Denny's arm lightly with her hand as she sashayed past him on her way to a large chaise lounge set in full sun by the pool. "I hope we get to see a lot more of one another over the next few days," she said in a suggestive tone.

Denny was at a complete loss for words. Had she been coming on to him? Why would a beautiful movie star flirt with an old man like him, someone she just met? He shrugged and looked back at Larry who was leaning back in his chair with a huge grin on his face.

"She's some piece of work isn't she?" He clapped his hands and laughed. "Don't worry, Mr. Carswell, she likes to mess with men who visit the ranch. You're just her type; she seems to prefer handsome, middle-aged, outdoorsy types. She's really harmless," he said dismissively, "but be forewarned, if she thinks you're interested, if you chase her, things might heat up. That's one cougar who likes to prowl."

Larry was thoroughly enjoying the show put on by Bella, and he was most definitely juiced. Either his drink had been heavy on the vodka, or it wasn't his first cocktail of the day. If he had to bet, Denny would definitely choose the second explanation. Denny sighed inwardly and glanced once more in Bella's direction. She'd removed her sun hat to reveal smooth, shoulder-length dark-blonde hair that she'd tied back in a ponytail to keep it from covering her skin. "She really is striking," he thought as her taut figure glistened in the early afternoon sunlight. He frowned at the sight of her surgically augmented breasts sticking straight up, nipples pointing to the sky. As he began to put things into perspective he lamented inwardly, *For a woman who seems to have everything, it's too bad she's so insecure.*

"Well, you must have some idea of how you want the land divided," he said turning to Larry. "Have you been thinking about carving out a specified acreage or do you simply want me to run a boundary through a specified area?" For some reason Denny suddenly wanted to get on with this job, get it done and leave. Something was troubling him, but he couldn't pin down what it was. Things just didn't feel right.

Alas, Denny's premonition was right on the money. Little did he know he was already in over his head.

IV

Denny and Larry exited the pool area and began walking around the ranch house to where Denny had parked his truck. A large barn with stables sat off to their right, partially hidden by several towering Ponderosa pines. At the sound of voices coming from the stables, Larry beckoned Denny toward their source. They detoured up a well-worn path covered with wood mulch, pine needles, and the occasional pile of dried horse manure, to a paddock where two individuals, an older man in his late fifties and a petite, dark-haired young woman, were engaged in conversation regarding one of the horses.

"Hola, Emilio," Larry called out as they approached, "I didn't expect to see you back so soon. Is everything all right back at the set?

Emilio handed the reins of the horse to the woman and spoke to her quietly before sauntering over to the paddock fence.

"One of the horses came up lame so I brought the stock back here to rest 'em. The next shot's not far from here, down by Abiquiu, so I don't see the point of leaving 'em there. 'Sides, they won't be ready to use 'em for at least two days; I'll drive a fresh set of horses down when it's time."

"Emilio runs the stable here at the ranch," Larry explained, "and he's also been asked to shuttle horses to the set of Dirk's new movie whenever they're needed. Oh, look at me," he clucked, "where are my manners? Denny, this is Emilio Sisneros." He looked at Emilio, "Denny is the surveyor who's going to plat the land we're selling to the pueblo." His countenance brightened as a thought came to mind, "Say, Emilio, Denny would probably appreciate a guided tour of the property. Would you mind showing him around?"

The look on Emilio's face was one of indifference. He shrugged, "Sure, why not? Lily can handle things here. So, Mr. Surveyor, you ever been on a horse?"

Denny was no cowboy, but he'd ridden enough to respond with a nod. "Just let me get a few things from my truck before we head out."

"Okay, pardner, I'll bring the horses around front and meet you there."

Denny and Emilio took off in different directions leaving Larry standing alone beside the corral fence. With a frown he began to walk back to the house on wobbly, inebriated legs, but suddenly stopped and turned to face the barn, "Good afternoon, Lily," he yelled with a wave. The young woman gave a slight tip of her hat and disappeared into the barn.

Denny pulled a notebook, a compass, and his hand-held GPS unit from behind the seat of his old Ford. He leaned against the fender of his truck and thought about the job at hand, trying to forget the strange nature of his conversation with Larry and Bella. There wasn't much to do except wait for Mr. Sisneros to show up with their mounts. Opening the folder given to him by Larry, he extracted the deed and re-read the description of the last boundary course . . . "Paraje San Cristobal." The phrase surprised him because it was a modern document. On some of the old land grants bestowed by representatives of the King of Spain in the seventeenth century, "parajes" were "stopping points," important landmarks referred to in those conveyances. Many bore the names of Catholic saints, and Denny had only run across one or two references to them in his forty-plus year career. He made a mental note to show the deed to his good friend and mentor, Bonifacio Romero, the next time they got together. The crunch of tires on gravel interrupted his thoughts. A car had pulled around the driveway and was rolling to a stop next to him.

It was an odd-looking car, a fire-engine red convertible with spoked wheels and a strange-looking emblem on the hood—some sort of foreign sports car. The driver extricated himself with some difficulty from the low-slung seat. His expensively tailored suit, designed to disguise his drooping shoulders and burgeoning pot-belly, had the opposite effect of drawing attention to a body in need of exercise. In like fashion, a comb-over, intended to hide his balding pate, flapped in the breeze. The fellow leaned back into the car and grabbed a leather attaché case from the passenger side. Denny was curious. "That's an interesting set of wheels you've got there. What make is it?"

"It's a Lamborghini. Don't tell me you've never seen a Lamborghini before."

"Well, I probably have, just never close up. It looks fast."

"Yeah, that it is," the man said with pride. "Go ahead and look it over, but don't touch anything; I just had it waxed."

Denny decided he wouldn't go anywhere near it. The fellow seemed overly possessive. Instead he decided to introduce himself, and in the process learned that his new acquaintance's name was Conrad Sussman and that he was the lawyer who would preside over the land sale related to his survey.

"Barring unforeseen problems, I'll do my best to have the survey of the property ready for the meeting on Thursday," Denny offered. "Once the parameters of the sale are agreed upon I'll only need another day or two to prepare the plats you'll need to accompany the deeds." He figured he'd be burning the midnight oil in order to finish this job on time, but he was in a hurry to get it over with. Besides, if all went well he'd be done by the weekend, and he and Pam had talked about doing some fishing and hiking in the Pecos wilderness, enjoying the stands of aspen that were turning to gold as the days continued to shorten. At the moment, Pam was at an architectural conference in Phoenix. Both she and Denny had been on the road for most of the last three weeks, and they were looking forward to spending some quality time together.

"That would be helpful," Conrad admitted. He paused for a moment and pointed at Denny's truck. "Will you be driving *that* to the realtor's office on Thursday? No offense, but this *is* a million dollar deal. I personally feel that certain standards should be adhered to when such business is conducted. Maybe you can borrow Dirk's Mercedes to drive to the meeting."

That Conrad Sussman was a pompous ass was plainly evident, and normally Denny would have shrugged off such helpful advice, smiled, and walked away. After all, a man is only as big as the things that upset him, he reasoned, and Conrad wasn't worth fretting over. However, after the morning's long drive to the ranch, and after the uncomfortable conversation he'd had with Bella and Larry, Denny was a bit frazzled. He let the arrogant comment about his truck get to him. Anger welled up in him and he was about to say something he might really regret when another vehicle came roaring up the driveway. This one was diametrically different from the one driven by Conrad. It was an ancient pick-up, even older than Denny's, an International Harvester dating from the late fifties or early sixties. There was a funny-looking homemade camper mounted in the bed, and in the driver's seat sat a young man with long blonde hair tucked under a strange-looking hat.

"I haven't seen a truck like that in decades," thought Denny as the driver skidded to a stop right next to the Lamborghini, kicking up rocks and dust in the process. The man in the cab just sat where he was and smiled at Conrad. From the look on the lawyer's face it was apparent the two weren't the best of friends.

The truck's arrival had given Denny time to settle down and collect his thoughts. He reconsidered his proposed statement to Conrad, one originally planned to include a reference to self-fornication, and toned it down several notches. "Despite your advice, Conrad, I *will* be driving to the meeting in this vehicle. It may not be worth a lot of money,"—he chuckled, "I like to tell folks that it doubles in value every time I fill the gas tank—but she's paid for, she's got four-wheel-drive, unlike many fancy sports cars, and she gets me and my equipment where I want to go safely and reliably."

Conrad adopted a dour expression and looked in turn at Denny and the man in the other truck before beginning to make his way to the ranch house. "Just stay away from my car," he muttered in passing.

"Dude!" said the fellow in the old International. He scrambled out and came over to shake Denny's hand. "That was awesome. I mean, like, I've never seen anyone give it to 'ol Suss Puss like that. Way to go man. By the way, my friends call me Xeno. What's your handle?"

Denny introduced himself and explained about the survey. Xeno seemed fascinated by what Denny had to say and immediately began interrogating him about the equipment he was going to use on the job. In particular he wanted to know all about the GPS unit Denny had and how the receivers handled the signals emitted by the satellites in the NAVSTAR constellation. The young man's knowledge was surprisingly extensive, but before their discussion became more involved, Emilio and Lily showed up leading two horses.

"It looks like my ride is here," Denny said, "but if you'd like to hang around when I run the survey you're perfectly welcome. It appears we have a lot to talk about."

EMILIO AND DENNY RODE slowly through ponderosa forest along well-maintained and heavily used trails. The soil was dry; there hadn't been much rain in the past month, and the horses stirred up dust in areas that would usually have been wet. Occasionally his host would point out a landmark to help Denny with his bearings. He'd studied the plat enough to have a pretty good idea of where they were and where they were headed, but it was still nice to have a guide; the ranch covered well over twelve hundred acres. Emilio was a quiet fellow and Denny was enjoying the peaceful scenery, but he finally decided he should initiate some sort of conversation.

"So, how long have you been working here on this ranch, Emilio?"

"Oh, let's see, 'bout eight years now I reckon'. I'm not new to the area though–lived around here most of my life."

"Eight years? Isn't that about how long Dirk and Bella have been married? Was Bella here when you hired on?"

"Yeah, in fact she's the one who likes to ride. 'Ol Dirk wouldn't be caught dead on a horse unless he's forced to do it for a movie or somethin'. I guess they, I mean, she, wanted to bring horses onto the ranch and they needed someone to take care of 'em.

I was available and my friend Bobby Wade recommended me for the job . . . ." He shrugged and the two of them rode once again in silence.

"I must say, this horse is well trained and well behaved. I'm no expert, but I've ridden worse."

At this last comment Emilio smiled, "Betsy there is Dirk's mount. I just ran her back from that movie he's makin'. She needed some exercise and I didn't know how competent a rider you were. I wanted you to be safe. But, yeah, she's real sweet, almost fourteen years old now and startin' to get lazy. When no one rides her she likes to stand around where it's boggy down at the bottom of her pasture. See, she likes the grass down there, but when she goes there her hooves get soaked and get real soft, gives her big 'ol paddle feet, slows her down. That's another reason I wanted to get her out on the trail today."

Talking about horses had put Emilio in a more outgoing mood. He gradually became ever more loquacious, revealing interesting details about the ranch and the Stetson household, even a bit more than Denny cared to know.

"This is a pretty good job really. Dirk and Bella have . . . I guess I should say Agatha and Woody have treated me good." He looked back at Denny and smiled, "Them's their real names. I don't know how common it is for actors to change their names; I know the 'Duke' did it, and there must be others, but you got to believe anyone wanting to be a cowboy movie actor would be pretty lucky to be born with a Christian name like 'Stetson.' He's changed it legally, but he was originally Woody Hertzfelter. Bella was born Agatha Cruz. My advice is don't call Dirk 'Woody' unless you're tryin' to get his goat. It pisses him off." He chuckled, "I do it occasionally just to see him get all flustered. The 'sum bitch got no sense of humor at all, not one lick." He sighed, "The poor man never knows when you're kiddin' him. He's dumber 'n dirt. Makes me feel bad sometimes, like I'm tormenting some defenseless animal."

"So, how did you find this out, about their names I mean?"

"Bella told me. She tends to talk a lot when there's, uh, passion involved."

Denny caught the hint of a smile on Emilio's face. "So, you and Bella . . . ?"

"You mean did we 'keep company'? Yeah, we sure did, yessir." He stopped to contemplate what he was saying and shook his head. "But that was a while back, years ago. She's what you might call a Will Rogers woman—you know, never met a man she didn't like. Most of 'em don't stay with her too long; she gets bored, moves on after playing 'em for a while. I don't know how many notches she's got on her gun, but I bet it's a bunch."

"I kind of got that impression earlier today," Denny admitted. "What I find sad is that she's so shallow, you know, having had all that plastic surgery. I think she'd be more beautiful if she didn't try to look half her age."

"You and me both. Her doctors must'a used a hundred tubes of Dow Corning silicone to make them titties stand up the way they do, but it just don't look natural; you got that right."

Denny was more than ready to change the subject. He turned his thoughts to the reason for his being there. On a hunch he decided to see if Emilio could enlighten him about the language in the deed; after all, he'd stated earlier that he was from the area. Maybe he'd absorbed some of the local lore regarding place names.

The trail meandered through a south-facing meadow. Brilliant sunlight illuminated the Rio Casteneda valley, and a cool breeze welcomed them into an open space with a spectacular view. Stands of aspen stood uphill from them, their leaves quaking and fluttering in the wind. Most had already turned a golden yellow. Both men, it seemed, were reluctant to speak while Mother Nature demanded their awe and attention on this fine fall day. When the trail entered a wide draw and encountered a canopy of pine, Denny questioned Emilio about what he'd read. "Did you ever know or hear of a man named Antonio Roybal? He surveyed this ranch back in 1995."

Emilio thought hard before responding. "No, can't say as I have. You might try askin' around in Arenas Rojas. Someone there might know something."

"What about the term 'paraje San Cristobal' or a place called Cerro Medio? It should be a hill on which one of the property corners is located."

"Well, Sir, you must have some of that 'extra special perception' or whatever it is, 'cause that's where we're headed. In fact this hill we're on now is the Cerro Medio. Heck, now that I think about it, some of the old timers call it San Cristobal hill. Could be that's why you saw it referred to both ways."

What Emilio said made a lot of sense, but the word "paraje" is what really intrigued Denny, especially given that it was used by a surveyor in a boundary description. He made a mental note to contact Mr. Roybal as soon as possible.

The trail grew steep and rocky as the two men rode up the last two switchbacks leading to the top of the hill. A wonderful panorama rewarded their arrival at the summit. The entire bed of the Rio Casteneda could be seen circling around them from south to west. It crossed the road as it neared the town of Arenas Rojas, and then disappeared as it wound behind tree-covered hills to the northeast.

Emilio dismounted and grabbed Betsy's reins indicating to Denny to do likewise. He led them to a pile of rocks containing a survey monument. Denny studied it, noting that it bore the name Joshua Marquez along with his New Mexico Survey registration number. A barbed wire fence ran down the hill to the southwest all the way to the arroyo Casteneda. Obviously it was the fence shown on Marquez's plat. The terrain to the west was much less steep than it was to the east, the side Denny and Emilio had just climbed.

"I saw Tommy Hawkfeather in town a few weeks ago," Emilio said. "He's a buddy of mine from Santa Teresa. What he told me is that the tribe wants all of the land from this fence right up to near the ranch entrance over there." He swept his hand across the view of the river valley. "All that low land is where they want to put their casino."

As he gazed upon the land indicated by Emilio, Denny bit his lip in concentration. Opening the side pocket of his backpack he extracted his compass and an old six-inch slide rule. The tract in question was bounded by the dry river bed which followed a nearly circular arc centered roughly at the point where they stood. Estimating the radius of that arc at just under half a mile, give or take, he noted the bearing along the fence and then the bearing which closed the arc, located off to the northeast in the approximate direction of the gate he'd driven through that morning. A few quick strokes on the slide rule yielded an area of approximately 180 acres, and at $7500 per that came to a cool $1.3 million, a helluva nice sum for undeveloped scrub land. He returned his tools to his pack and was about to ask Emilio something when a dust cloud and engine noise distracted him. Climbing the Cerro Medio from the southwest was an old light-green pickup negotiating its way along an overgrown two-track path. *Someone's in a bit of a hurry,* he thought as he watched the driver bounce around the cab, driving too fast over the bumpy terrain. Brakes squealed and the truck skidded to a stop a few yards away.

"Shit," muttered Emilio. The cab door slammed shut and an intense, slightly-built older man with white hair and a limp approached as fast as his legs would allow, breathing hard from the effort.

"Come on, Augusto," Emilio groaned, "don't be bustin' my balls today, okay? We're not trespassin'."

"Yeah, you are. I just happened to be workin' in my barn and saw y'all ridin' over the hill," he wheezed. "What I'd like to hear is you tellin' me that 'ol dumbass Dirk finally saw the error of his ways and sent you to take down that fence."

"Ain't gonna happen, Augusto. You've seen the paperwork. That last property survey put the matter to rest. The judge agreed. In fact, this here gentleman is another surveyor. The Stetsons plan to sell this land to the pueblo and he's got to stake it out."

Augusto Benavides settled down and caught his breath. When he continued there was disgust in his voice.

"Yeah, that's what that traitor nephew of mine keeps tellin' me. That cabrón lawyer sold out his family, sold his soul to make a buck by brokering that deal. Well, it ain't the first time someone's tried to cheat this family out of its rightful land. But I'm tellin' you

'Milio, I won't let it happen without a fight. I'm gonna get me another lawyer and maybe get me a hired-gun surveyor to block that sale." Glowering at the others, Augusto waved them off and hiked back to his truck.

Denny stood dumbstruck at this latest development. An angry adjoiner was, in his mind, the worst complicating factor to be encountered on a survey job, and during his forty-plus year career he'd seen just about everything: bad and erroneous monumentation; an innumerable array of documentation problems such as ambiguous and poorly executed conveyances, deeds and easements; hazards such as wild animals, steep, unforgiving and dangerous terrain, etc. But the one thing he dreaded the most was the irate neighbor. Humans were clever, unpredictable creatures who, if they decided to cause trouble, could be a serious threat to one's job, livelihood, and perhaps to one's life.

Emilio noticed the concern on Denny's face. "Aw, don't worry about Augusto; he's harmless really. We're pretty much friends, but when he gets riled up about his land he can be a pain in the ass. Dirk and Bella had to get that surveyor in here three years ago to shut him up. That fellow, what was his name . . . ?"

"Joshua Marquez," Denny interjected.

"Yeah, him. Nice young fella. Anyway he had one heck of a time gettin' his job done. Finally the whole matter went before a judge where all the evidence got presented. Augusto and Maria—that's Augusto's wife—had their say, but I guess the deed says the land belongs to the Stetsons. Case closed."

The Sun dove behind clouds and the air took on a chill, reminding Denny that winter was indeed just around the corner. The sudden cold, foreboding atmosphere mirrored the change in his mood. Something Augusto Benavides had said was nagging at him, but he couldn't organize his thoughts well enough to clarify what it was. There was more to the history of the Stetson ranch than met the eye. Although it pained him to think about it, he knew he would have to talk at length with Conrad and perhaps pay a visit to Augusto in order to figure out what was going on. First and foremost, however, he had to spend some time in the county offices perusing the deed histories of both tracts, and he definitely had to talk to the surveyors who had worked on them recently. Already this job was getting complicated, and Denny had the sinking feeling that, in the end, someone's feelings were going to be terribly hurt when the truth came out.

He was ready to leave, but at the sight of Emilio lighting a cigarette, he figured he'd give his escort a break before they headed back. Several lichen-covered rocks at the cairn surrounding Joshua Marquez's boundary marker provided a convenient place for him to sit and relax while he waited. Looking out over the steep-sided slope to the northeast he could see the front porch of the ranch house. Beyond that he could make out the horse barn and the pasture that led to dense ponderosa and fir forest. Back by what he knew was the other side of the property, a rock outcrop arose from the woods yielding a small knoll. From that vantage point and from the top of Cerro Medio, there was a good chance he'd be able to see most, and perhaps all, of the corners of the property, a good sign relating to the field work portion of the job.

Movement in the parking area caught his eye. The red sports car belonging to Conrad was moving down the driveway. Soon the car emerged from behind the tree line that obscured the main gate. Denny absentmindedly followed the car as it turned west onto the highway and moved across his field of view, a blood-red spot followed by a grey cloud of dust. Sunlight glinted off the chromed grill and side mirrors, and Denny could just make out Conrad's hair flopping along one side of his designer sunglasses. He was about to get up and talk to Emilio when something else caught his eye, an object that was out of place in the scrub grassland lying between him and the Lamborghini. An old fencepost stuck out of the ground at an angle, and as he continued to stare, several more came into view. Some had segments of heavily rusted barbed wire tacked to them. The fence had been largely removed, but there was clear evidence that it had once run from the corner where he sat, northwest towards the highway and the town of Arenas Rojas. Another warning bell went off in Denny's head. Fences often constitute strong evidence of property lines. He would have to study all documents of record related to this job very carefully; that fence almost certainly meant something.

The Lamborghini was now a speck in the distance. Denny stood up. "Emilio, what can you tell me about the lawyer, Sussman?"

Without looking at Denny, Emilio dropped his cigarette butt into the dirt and stubbed it out with the toe of his boot. Several seconds elapsed making Denny think he hadn't been heard. He was about to repeat the question when Emilio replied in a low voice, "Don't know much about 'im 'cept that he's a prick. That's something both Augusto and I can agree on." Emilio continued to look away while he re-mounted his horse. The touch of venom in his voice indicated to Denny that the subject of Conrad Sussman was probably best left alone.

Denny climbed aboard Betsy and the two began to ride in silence along the well-worn path bordering the newer fence line to the west, toward the Rio Casteneda arroyo. A gentle breeze pressed the back of Denny's shirt into the line of perspiration along his spine; the evaporative coolness, bordering on chill, provided a welcome contrast to the afternoon sun that warmed his face. Emilio was riding slowly, evidently in no hurry to get back to the ranch. The lazy pace provided ample time for further conversation. When the path widened, Denny eased Betsy alongside Emilio, "I met an interesting young man in the parking lot just before you showed up with our mounts."

Emilio laughed, "You mean Xeno? Yeah, he's definitely different, sharp as a tack and crazy as a loon. Lily knows him better than I do; in fact, I think she's sweet on him. Anyway, she tells me he was studying to be some kind of rocket scientist a few years back; I think it was at the University of New Mexico. One day he just up and quit, walked away from physics to live in his truck and worry about little green men from outer space."

"He did seem a bit odd, but he's incredibly knowledgeable about the GPS satellite constellation. He wants to tag along while I survey the property; he seemed excited to learn more about the receivers I use."

"I'm sure he is if it has anything to do with space—thinks he can communicate with men from Mars. If I were you I'd keep a close eye on my equipment with him around. You just never know."

"What's he doing way out here on the Stetson ranch?"

"Oh, he didn't come out here 'cause of the Stetson's. He hangs around over on Augusto's spread mostly, says it's the perfect place to meet spacemen. 'Gusto lets him camp in the woods somewhere and Xeno helps out with chores on the ranch. He's been there better'n a year now. Last spring he was workin' fence line when Bella spotted him while she was out ridin'." He chuckled and looked at Denny, "I guess she liked what she saw; he must'a had his shirt off or somethin'. Anyway, she gives him jobs on our ranch now, mostly around the pool. I see her watchin' him while she sunbathes—takin' in the view."

The Sun was low in the sky when Emilio and Denny got back to the stables. Denny walked gingerly as he carried Betsy's saddle to the tack room. His knees were sore and stiff from the long ride. "Now I know why cowboys are so bowlegged," he groaned. He thanked Emilio profusely for the tour of the property even though he suspected that the ride had been a godsend for his host, a welcome reprieve from his daily grind. With his tattered backpack slung over his shoulder, Denny began to walk back to the front parking lot. It was late, after four PM, probably too late to swing by the county offices, but there was still plenty of time to get hold of Marquez and Roybal, the two men who most recently surveyed the property.

The route to his truck brought Denny past the patio and pool where several people were talking animatedly. Bella was there, sharply but casually dressed in sandals and calf-length khaki pants. Her hair was no longer tied back; it hung loosely and stylishly at her shoulders, brushing the collar of a loose-fitting cotton shirt whenever she turned her head. She was engaged in conversation with both Larry and a bald-headed man who appeared to be about Denny's age. As he walked past, laughter broke out and snippets of what was being said floated Denny's way through the cool afternoon air. Larry, drink in hand, reacted loudly and drunkenly to something the other man said, "Ha, ha! Oh Andre, really? I can't believe he said that . . . ."

Denny smiled and kept walking. Their conversation was none of his business, but out of the corner of his eye he happened to notice the bald man, Andre, place his hand in the small of Bella's back and lean in to give her a quick kiss on her cheek. An unusual feeling swept through Denny at the sight, a strange mixture of embarrassment and something else—jealousy? Had the fact that Bella flirted with him made him feel a bit possessive when it came to her attentions? Denny prided himself on his objectivity. *Yes*, he thought, *I'm jealous!* He had absolutely no intention of involving himself in any way with Bella, so why in heaven's name should he feel the least bit possessive of her? Beautiful women can be very manipulative, he realized with genuine awe, even when they don't intend to be.

When he reached his truck, Denny heaved a sigh of relief. "Maybe I've been too judgmental," he concluded. In this stress-free moment he paused to reflect upon the afternoon's events. After his long talk with Emilio and the relaxing ride through the property, and after seeing Larry and Bella quietly relaxing out by the pool with friends, he realized he may have made too much of the encounter with a semi-nude Bella. He did, after all, intrude on the private lives of this group of people, invading upon their daily affairs through his visit. And didn't Bella point out that, had she known he was coming over, she'd have made herself presentable? Perhaps the Stetson's and their world weren't so strange after all. Perhaps this would simply turn into a routine but refreshingly quirky job that he'd be able to laugh about in retrospect.

Yeah . . . maybe.

V

The county clerk's office was indeed closed when he phoned at around 4:30. The recording that answered his call suggested they remained open until five, but Denny felt a degree of relief to find no one home. He was tired and hungry and he figured he might have aggravated the good public servants of Taos County by showing up close to quitting time. Experience had taught him that it never paid to antagonize anyone who presided over information one needed. Tomorrow morning he'd drive into Taos to perform the one or two hours' worth of research the job would require.

A growling stomach reminded Denny that he hadn't eaten since six AM. It took all of five minutes to drive through the commercial district of Arenas Rojas—twice—in search of culinary establishments. Downtown was laid out in a quadrangle with its small church, topped with a weathered cross, as centerpiece. The road leading up from the highway came in from the South, ending at a T-junction with a road forming the southern boundary of the central square. Between the road and the church was a small park, nicely xeriscaped with native trees that overhung cast-iron benches, and with a statue of the Madonna standing on an elevated pedestal in the center of a dry fountain. Commercial buildings lined the streets east and west of the church and included a farm and feed supply store, a hardware store, a unisex hairstyling establishment, and a small bodega/gas station. The town was so small that no fast-food franchises had yet established footholds, a fact he found decidedly encouraging. But he also discovered that his choice of restaurants totaled exactly . . . one: a small diner named 'Alice's Restaurant.'

Denny parked alongside the diner at the end of a line of vehicles, all of them pickups. He tucked in his shirt, ran a comb through his hair and opened the rickety door to the establishment. A bell announced his presence. An overhead heater squeaked and blew warm air in his face while he took a moment to examine his surroundings.

"Take a seat anywhere, Hon."

It took Denny several seconds to figure out where the voice had come from: a waitress who had her back to him, cleaning a table. When she turned around she wore a big smile, "If it's just you, have a seat at the counter." She hurried past with an armload of dirty dishes, "You want a menu?"

"Please," he replied as he eased himself onto a worn vinyl-covered stool, out of range of the heater.

The name of the place had planted a song in his head. He caught himself whistling the namesake Arlo Guthrie tune when the waitress returned, brushing an errant wisp of hair from her eyes. "Yes, Sweetie, you can get anything you want in Alice's Restaurant," she smiled, "as long as it's on the menu."

She was a jovial, stocky woman, seemingly in her mid-thirties, possessing a million dollar smile, a nametag that read 'Jenny,' and an obvious upbeat outlook on life. Denny took an instant liking to her. He'd have to return to eat here again, assuming the food was palatable. After brief study of the dinner list he settled on a chicken and black bean enchilada and a Sprite.

"Red or green?"

"Chili verde, por favor."

She giggled, "You got it, handsome. It'll be ready in about five minutes."

Inspiration hit him, "Say, ma'am, I've come up from Santa Fé to help with a land survey and I was wondering: do you know a man named Joshua Marquez? He did the last survey of the property in question." It was a lot of information to give her, but because he was from out-of-town he figured he owed anyone local an explanation of why he was here, especially if he was going to enquire about one of their own.

"Well, you picked the right person to ask. Josh and I went to school together."

"Order up." The authoritative masculine voice was that of the cook who'd just placed two plates on a stainless-steel pass-through in back of the counter. Jenny held up a finger to indicate she'd be right back and hurried off to deliver the dinners to two women seated at a booth. When she returned she picked up where she'd left off.

"Josh Marquez. I don't see him very often these days. He moved to Farmington a few years ago. There was more business for him out that way." She sounded wistful, as though he might be an old flame. "His grandmother, Lucinda, still lives here in town; if you look her up I'm sure she can put you in touch with him." She bustled away to wait tables and later, as Denny was mopping up the remains of chili, cheese and enchilada sauce with a bit of sopaipilla, Jenny handed him his dinner check. Coupled with the check was a blank receipt bearing Mrs. Marquez's phone number. Denny thanked her and left a generous tip.

The sky was pitch-black because of a waning moon. Despite the light emanating from the diner's neon sign, the constellations of autumn stood out against an inky background. Without dark-adapted eyes, Denny had trouble locating the door handle to his truck, but just as he grasped it he was startled at the mention of his name.

"Mr. Carswell?" He turned, expecting to see one of the people he'd met at the Stetson ranch, but to his surprise the speaker was someone he'd never seen before. The fellow was short, with a wiry build, and sported a wide-brimmed felt hat with a large feather attached. The man smiled, "I'm sorry if we've interrupted your dinner." He turned and nodded to his friend, a heavyset man with a long braided ponytail, who, for whatever reason, chose to keep his distance. "My name's Tommy Hawkfeather. Emilio Sisneros told me you're surveying the property we're trying to acquire. Also, you've probably been told how important this land sale is to us. I just wanted to introduce myself and let you know that if we at Santa Teresa Pueblo can be of any assistance, please don't hesitate to ask." They shook hands and with a tip of his hat, Hawkfeather, with his large companion in tow, disappeared into Alice's restaurant.

Denny stared after them not knowing just what to make of the brief encounter. He remembered Emilio mentioning the fellow's name and referring to him as a friend. Hawkfeather certainly seemed congenial enough, and maybe he was just being friendly, offering to help if needed. But there also seemed to be a subtle undertone to his message, something mildly intimidating. Forcing paranoia aside, Denny got into his truck. A glance at his watch gave him pause. It was definitely not too late to call Lucinda Marquez. Pulling his cell phone from its holster and the paper with her phone number from his shirt pocket, he punched in her number. Mrs. Marquez answered immediately.

Her high-pitched singsong voice suggested she was a petite woman, and her response was fast and alert, not slowed by age as it might have been. After introducing himself and stating the reason for his call, he fell silent as Mrs. Marquez launched into an almost reverent account of Josh's life. This opportunity to talk about her grandson would not be squandered. Denny listened patiently, at first out of courtesy, inwardly fretting about the evaporation of so many precious cell phone minutes, but he gradually became interested in the individual who had performed a job closely related to what he himself was about to undertake. Did he know that Josh had been valedictorian at Taos High? Captain of the track team? Josh had been no ordinary student athlete. He'd been a distinguished champion in the 800 meter run, a grueling two laps of an Olympic-sized track. A blazing time of one minute fifty-one point four seconds in the state high school finals in his senior year (Lucinda remembered the time to the tenth of a second) had earned him a partial scholarship at New Mexico State University. He'd made good use of his athletic skills, propelling himself, over four years, to his bachelor's degree in Surveying Engineering, cum laude. Listening to this proud grandmother talk about her grandson made Denny think of the numerous parallels between Josh Marquez and Greg Martinez, Denny's apprentice. Greg was also the product of a single parent household, a hard-working lad who was putting himself through the same academic program by working with Denny during summers and academic breaks.

When he finally sensed an opening in the conversation, Denny interjected a query about how to contact Josh. 'Trask Engineering' in Farmington, New Mexico was where he now worked. Lucinda recited both his work and cell numbers from memory.

DENNY CARSWELL'S CIRCADIAN CLOCK never let him sleep past 6:00 AM. He was a morning person through-and-through, someone who always enjoyed an enhanced period of productivity early in the day. Normally. The work he had planned for today, however, depended upon others, so Denny had an hour or so to kill before getting started. Rather than remaining cooped-up in his hotel room, he opted for coffee and danish at the Dairy Queen down the street.

Even at that early hour, Denny was not at a loss for company. Members of the local community drifted in and out continually, stopping to chat and fuel themselves before the start of another day. This was, evidently, a hot spot for the morning crowd. Most seemed to know one another. "Regulars" Denny guessed, who craved a bit of relaxed social camaraderie as a prelude to their working days. The fact that he was from out of town didn't matter to those who sat beside him. Two congenial young men, wearing the clean, pressed uniforms of a local auto parts store, chatted with him briefly while their standing order of coffee and a large box of doughnuts was made ready. A loquacious postmistress entertained everyone in attendance by recounting the recent trip of her three-year-old to the barber shop. Several delivery men, some of whom had been on the road since long before Denny had woken up, paid homage to the java deity in brief respite from their daily pilgrimages around the State.

Denny thoroughly enjoyed these relaxed interactions. He thought about how nice it would be to become a "regular" at some coffee shop near home, to go where 'everybody knows your name.' Alas, it didn't fit with his personality. He always put internal pressure on himself to get an early start to his workday. His conscience would nag at him unmercifully to get his ass in gear if he were to engage in such an idle morning routine no matter how brief it might be. Speaking of which, he consulted his watch—7:30. Given what Mrs. Marquez had said about her grandson, he figured it wasn't too early to ring him up. He was right.

"Hello, Mr. Marquez? My name is Denny Carswell. I'm a surveyor. I normally work out of Santa Fe, but today I'm in Taos trying to . . . ."

"Carswell? Are you the same Dennis Carswell who helped solve that murder down near Silver City?"

"You heard about that, huh?"

"'Course I have. You're a bit of a celebrity, Mr. Carswell, at least in the surveying community."

Denny wasn't exactly thrilled with the attention he'd garnered from that awful experience back in May. He shifted the subject back to the task at hand. "First of all, it's 'Denny,' not 'Mr. Carswell.' Secondly, I have a few questions to ask about a job I've acquired near your home town of Arenas Rojas. There's a ranch here owned by a Mr. Dirk Stetson. He wants to parcel off a portion to sell to the Santa Teresa pueblo. You performed the last survey of record on the property."

There was silence at Josh's end. The mention of the Stetson ranch had given him pause. "Yeah, I remember that job. What about it?"

"My questions don't relate so much to your survey—that was a very clean and professional bit of work, by the way. What I'm after is information about the job done by Antonio Roybal back in 1995. The verbiage in his property description seems a bit strange. The few people I've asked about him haven't been any help and I don't know how to contact him." Denny's hotel room had WiFi which he had exploited, to no avail, in attempts to locate Mr. Roybal.

"The probable reason why nobody knows about him is because he's been dead since soon after he did that job, a heart attack as I recall. He also wasn't local; I think he worked out of an office in Española. You're right about that deed though. Part of his metes and bounds description seemed incomplete. Oh, and the document didn't contain a plat. I had to rely on what he'd written, interpret it as best I could, in order to do the job. The description of the adjoining property, the Benavides ranch, wasn't any help either. That survey turned into a monster headache for me and for all parties concerned. Mr. Benavides contested my findings and the whole mess went to court."

"Darn," Denny muttered, "I'd planned to spend the morning in the Taos County clerk's office to learn more about these properties. From what you've said I'll probably be wasting my time."

"I hate to say it, Mr. Carswell, er . . . Denny, but I think that's the case. However, it's entirely possible that I overlooked something; I'm certainly not infallible." He paused again, unsure of whether or not he should continue. "For what it's worth, the ranch manager, his name is Larry something, called me and asked if I would do this job, but after the trouble I had with that neighbor, Benavides, I turned him down flat. Tell you what, give me your phone number and if I think of something that may help you I'll call. Other than that the best I can do is to wish you good luck."

Denny thanked the young man and hung up. *I'd still better check the records office,* he thought, *even if it's probably a dead end.* His spirits were dropping. The uneasy premonition that had come over him the day before was becoming manifest. This job would most likely get worse before it got better. He ordered another coffee to go, paid his bill, and headed to the Taos County office complex.

DENNY WAS IN A DESPONDENT mood as he drove through the main gate and entered the Stetson property. Josh Marquez had been right. The only part of the Stetson ranch that bordered private land lay to the south and west, adjacent to the Benavides property. The northern boundary was the highway, and to the east lay National Forest. What he hadn't appreciated until this morning, however, was that the federal land was in Mora County. In fact, the Taos/Mora county line ran almost due north-south along the eastern edge of both Dirk's ranch and the Benavides spread.

The deed history of both private parcels was also sketchy and of little value. After the Mexican war in the 1840s, the US government took possession of all the lands in this area save for the acreage allotted to the Santa Teresa pueblo. The forest service subsequently sold its holdings west of Mora County and south of the State highway in 1905. Most of that land was purchased by ancestors of Maria Benavides and some by another private landholder further to the south. Augusto and Maria had then sold part of their land to Stetson in 1995. The only extant description of what they sold is Antonio Roybal's survey. Josh Marquez had had nothing else to work with.

The parking area in front of the ranch house was buzzing with activity. Some sort of party or get-together was in progress. More than a dozen cars were present along with a few vans from "Daylight Pictures." The company's logo was that of a brilliant sun from which orange rays slanted downward towards a far off tropical island topped with a solitary palm tree. Several people were milling around near the front door of the house, drinks in hand, relaxing and chatting amiably. The door was propped open and he could see others standing in the great room who were also enjoying themselves. A Muzak version of the Bee Gee's "You Should be Dancin'" drifted all the way out from the pool area.

Denny was anxious to start the field work part of his job. He'd lost nearly the entire morning messing around at the County offices "chasing windmills" as he was wont to say. It was time to get something done. There was a rough dirt road leading through the eastern part of the property along which he could drive most of the way to San Cristobal hill. He would begin the survey by setting his GPS base receiver at the paraje to collect satellite data, from which he could extract exact geodetic coordinates of that corner on a suitable datum. While that was in progress, a process that would take a few hours given his equipment limitations, he'd scout the ranch to determine the best way to access and pin down the locations of the other property corners. As he threw his truck into gear he noticed the quirky young man named Xeno sitting at the other end of the parking lot. Recalling their conversation of the day before, he felt obliged, out of courtesy, to invite him along while he set up his receiver. He shut off the engine and walked over to visit, noticing as he approached that Xeno wasn't alone; he was engaged in conversation with Emilio's assistant, the pretty young woman named Lily.

Xeno greeted him warmly. "Hey, Mr. Surveyor Dude, Que Pasa?"

"Hi, Xeno. I'm about to set up my GPS receiver. You had expressed an interest in assisting me?"

"Yeah, man, far out, but I'm afraid I've got to take a rain check. Gotta prepare for a big trip." He looked over at Lily who was seated on the ground chewing on a piece of straw, her back against a small Ponderosa. "I can't forget my manners. This is Lily, resident ranch scholar and horseback rider extraordinaire." The woman blushed and smiled. "And, Lily, this is, um..."

"Call me Denny."

"Will do Deneroo."

"Hello, Denny," Lily said. "Mucho Gusto."

"Likewise," Denny smiled. He turned to Xeno. "Well, if you're not going to be away for too long, you're welcome to tag along some other time. Just look for my truck; I shouldn't be far away. By the way, uh, what's going on around here; someone having a birthday?"

"Naw," Xeno chuckled, "Dirk, the man, invited some of his movie star friends to visit his ranch. They're filming the next scene for his blockbuster movie 'Conestoga Highway' down by Abiquiu. It's my guess most of the cast gets the day off while the worker bees, you know, the set crew, prepare to start filming."

"The vainglorious stars with their bright, shiny nimbi, unencumbered by the travails of the lower castes," Lily offered as she stood and slapped dust and dirt from her pants using her hat. She returned it to her head and sighed, "Oh, Dirk's all right. He just seems to be oblivious to most of what goes on during filming. When Emilio and I go to work on the set I'm a bit bothered by the condescending attitudes of some of the other actors towards those who work behind the scenes. That's normal I suppose; humans tend to form cliques and hierarchies in most group work environments; that sort of behavior was really bad in the restaurant where I used to work. Andre Bivens, the director of the film, isn't like that at all, though. He and his staff are admirably appreciative of the crew they've got. That's why I'm surprised he's here. It's out of character for him. You know what I think?" she smiled suspiciously, "I think he has a thing for Dirk's wife, Bella."

"Ha, ha, aren't you the gossip," Xeno laughed. "You're just trying to change the subject. What I was talking about earlier is important stuff." He winked at Denny before continuing. "That brother of yours should listen to old Xeno if he knows what's good for him. Let me see your hat." She rolled her eyes and reluctantly handed it to him. He, in turn, passed it on to Denny. "This here chapeau was expertly hand-crafted by Lily's brother, Carl."

Denny examined it closely. Made of high-quality felt, it had a turquoise-studded leather band with a matching drawstring and a chin-fastener made of tooled elk antler. A turkey feather stuck out of the band at an artistic angle. It was beautifully made, both functional and stylish, something Denny himself would be proud to wear. "Nice," was all he could think to say as he handed it back.

"I know, Dude," Xeno agreed. "And Carl makes other stuff too; bracelets, neclaces, belts—and I'm talkin' high quality ganga here, made with the finest leather, silver, turquoise—I keep telling his little sister that if big bro' wants to make real money, he has to learn the vagaries of 'marketing.'"

"Will you stop with your pecuniary maunderings?" she said in exasperated reply. "He's happy, Xeno. Leave him be. Carl considers himself an artist, not a merchant."

She's feisty, thought Denny with a chuckle. But he could tell that her outwardly irascible attitude was tempered, more bluff than real. He got the impression this sort of banter was normal for these two, a sort of verbal tug-of-war that both found entertaining. A relationship seemed to be in the offing, Emilio had expressed that opinion as an aside the day before, but he'd based his opinion on Lily's behavior alone. Denny could see that the converse was also true; Xeno had an interest in this pretty cowgirl, his verbal rival.

Lily was full-blooded Native American, with flawless bronze skin and a pretty, round face. Young and with an athleticism borne of hard work, she was petite and wiry with dark, expressive eyes. When she smiled in the sly fashion she was doing now, it had the effect of stirring a man's romantic interests. Xeno was completely under her spell as he continued his criticism of Carl's sales practices, now more for Denny's benefit than Lily's.

"I keep telling him, 'Carl, you can't sell art that way.' See, he thinks rich Tejanos, uh, Texans, you know, the guys who pull their 40 foot travel trailers with their spanking-new F-350s with full leather interiors? Well, he thinks when they drive through town they're gonna notice his "quaint" sign and say to their trophy wives, 'Gee, Mary Jo, what say we stop at the WalMart and get some more beef jerky and then go buy some art at that shop we saw down the road?'

"I told him, 'Look, if you want to make a living selling art, it's all in the way you present it. You make USDA grade 'A' stuff, my man, but it won't sell unless you turn your place into a tourist trap. You got to come up with a name like 'Spirit Water Cloud Gallery' or 'High Desert Butte Boutique,' and grow yourself a ponytail, or get a fake one, it don't matter; then you got to burn some incense and talk funny when they come to check you out. Oh, and you got to charge five times more than what you are; if you don't charge 'em 'til it hurts, they won't think they're getting high-quality merchandise. In fact, the actual quality don't mean shit, amigo, it's *perception* and *marketing* that count. Hell, most of them rich fuckers'll buy bronzed road apples if they're presented right. And the most important thing to remember: it's always the ladies who'll make the decision to buy something. The guys, well, the guys will spend whatever they think it'll take to get laid. Trust your buddy, Xeno, dude. I know what makes the world go 'round.'" Satisfied that he'd enlightened his colleagues with unimpeachable wisdom, Xeno crossed his arms and nodded with a sense of finality.

Denny had trouble suppressing a smile. He couldn't think of a response to the man's convincing argument, but he was curious about one conspicuous article of clothing that he wore.

"Say, Xeno, given that Carl makes such great headpieces, why do you wear . . . that?" He pointed to the unusual hat he'd noticed yesterday. He hadn't been comfortable mentioning it earlier, but it fascinated him. It looked like an old leather motorcycle or football helmet. Padded flaps with holes in them covered his ears and a silver band extended below the leather and down to his neckline. Xeno removed it and Denny could see that the silver layer was, in fact, aluminum foil pressed into the head cavity.

"This is my Faraday cage," he proudly remarked. "Keeps out all the electromagnetic junk society has been polluting the environment with for the past hundred years." He replaced the hat and wiggled it into place. "Also, keeps anyone from monitoring my thoughts. Tell me, Deneroo, does anyone really know for a fact that the government doesn't have long-range magnetoencephalographic systems for just that purpose, to eavesdrop on our most personal musings, checking for subversive tendencies, stealing novel ideas? My philosophy is: don't take chances, take precautions. It's better to be safe than sorry, my friend. No one's gonna fuck with Xeno's brain, not the domestic military-industrialists, and not the extraterrestrials. Nadia!" He grinned, "It helps attenuate the 'hum' too, when it gets real bad."

"The hum?"

"Yeah, man. You mean to tell me you've never heard of the 'Taos hum'?"

"Actually, I have," Denny replied, now genuinely intrigued by what the fellow was saying. "I just, well, I haven't yet met anyone who claims to have heard it. Is it real?"

"It's as real as the ground you're standing on, amigo, a real low frequency buzz, sometimes with a slow superimposed beat frequency of one or two Hertz. Not everyone can hear it. I keep telling Lily to make herself a Faraday cage like mine; it cuts out all the rf noise along with most of the hum, but, if you listen real hard, you can still pick up residual signal."

Again, Denny was at a loss for words. An awkward silence ensued before Xeno shouldered a backpack that had been lying at his feet. "Muchachos y Muchachas," he announced, "I gotta run. ET waits for no man." He winked at Lily by way of saying "Goodbye" and began walking purposefully to where he'd parked his old International pick-up.

"I wonder how much of that is real and how much of it's an act," Denny muttered.

"That's a good question," said Lily. "I probably know Xeno better than anyone, and he does enjoy messing with people's minds. On the other hand most of what he says . . . ."

She was interrupted by a shout from across the lot. Xeno was peering into the cab of Denny's truck. "Oh, man, you got an old Trimble unit. Cool!" The statement was in reference to the GPS receiver sitting on the passenger seat, the one Denny hoped to set up in the next hour or so. "Deneroo, you're gonna have to show me what that baby can do when I get back." With that, Xeno continued his trek towards parts unknown.

"You think he's pretty bizarre don't you?" Lily asked.

"Uh, yeah," he replied, conveying both sarcasm and understatement.

A huge smile spread across Lily's face. "What may surprise you is that he's borderline genius, at least as far as I can tell." Denny's face took on a puzzled look. "Before you interrupted us—and I don't mean that to sound like I'm offended—we'd just finished a spirited discussion on the mathematical legitimacy of using the Laplace transformation to solve certain types of higher order differential equations." She chuckled, "I claimed, as an aspiring mathematician, that the application lacks true mathematical rigor. He countered, from a physicist's point of view, that, as long as one gets the right answer, what does it matter?" She paused. "Xeno's been a great friend to me. He's provided me with impassioned encouragement as I've worked toward my degrees. He's helped me to focus on things that are truly important . . . ." She trailed off realizing that her tone was hinting at romantic feelings she harbored for her strange friend, something too personal to share.

"Degrees . . . plural?" Denny asked.

"I'm trying for a double major at the University of New Mexico branch campus over in Taos: Mathematics and Classic American Literature."

"That's a strange combination," Denny offered, "and a double major—that sounds like twice the work."

"A lot of my friends say the same thing, but to tell the truth, my interests are so multifarious that I had trouble narrowing my coursework to just two areas. The math courses are frustrating, not because they're difficult but because my professors now have to send videos of their lectures from Albuquerque. All of the advanced courses are taught only at the main campus. Sometimes things get delayed in the mail. I'll often get two or three tapes at once, so I have to cram to keep up."

"That must be a tough way to study mathematics," Denny said. "What do you do when you need help?"

"Well, to be honest with you I've never had any problems with the subject matter," she replied somewhat embarrassed. "Xeno says I have a 'perspicacious mathematical mind'." She shrugged, "He may be right. Certain so-called abstruse concepts seem rather clear to me. I don't mean to sound immodest, it's just that, well, math just makes a lot of sense." She looked down at her watch, "I have to get back to the stables. It was nice conversing with you."

"I don't want to detain you much longer," Denny said, "but I have one more question. Where is Xeno going?"

"He didn't tell you?" She smiled, "Every once in a while he makes a pilgrimage to the Gas Buggy site."

"Gas Buggy?"

"I really have to be going." There was now mischief in her eye. "You wouldn't believe the explanation coming from me anyway. Xeno will have to fill you in when he gets back."

A dumbfounded Denny Carswell watched her climb the path that led to the horse paddock. The discourse he'd had with those two young people was one of the strangest he'd ever had, right down to Lily's last cryptic statement. As he slid into the cab of his truck, he mumbled to himself, "*No matter what else happens on this job, nothing will surprise me.*"

In less than 48 hours Denny Carswell would eat those words.

VI

At the paraje Denny disconnected the cable from his GPS battery, wound it up, and stuffed it into the Velcro-sealed pouch in the case with his GPS receiver. It was almost 5 PM now. He worked amongst long shadows spreading eastward from his truck and tripod as the burnt-orange Sun sank towards the distant hills. Data collection had gone without a hitch. After dinner he would post-process the RINEX files stored in the receiver's memory and fix the coordinates of Josh Marquez's monument, with 95% confidence, to within a centimeter. Black magic is was it was—pure and simple. He gazed up at the clear blue autumn sky and marveled at what he knew to be there but couldn't see—the eight satellites orbiting 11,000 miles above the Earth's surface that had beamed down the information now stored in his GPS receiver. Denny's spirits were high. He knew today's field results would be excellent, and while the equipment here at San Cristobal Hill had chugged away processing satellite signals, he'd had opportunity to scout around the ranch and plan what he would do tomorrow morning.

Once everything was safely stowed behind the passenger seat, Denny shut the door and leaned against the truck's fender. Today had been unseasonably warm, capping nearly a week of the kind of Indian summer that only happens once every few years. He rested quietly, reveling in the serenity, isolation and vistas the site provided. Aspen leaves applauded the passing of a faint, warm gust of wind. A raven clucked a message to a colleague before the raucous beat of its wings signaled its relocation to a more suitable venue. He listened hard and was disappointed to hear the steady drone of a small aircraft filtering in from the distance. In all of his 65 years Denny had been in only a handful of places where no sounds of human activity could be heard. To him, sound was perhaps the best indicator of human influence on the environment. He had a habit of pausing and listening intently whenever he was in a secluded place, to test its level of privacy and solitude. As he concentrated, and his hearing became more acute, he could now discern motor noise wafting up from the streets of Arenas Rojas and faint strains of music coming from Dirk Stetson's party. With a sigh he pushed himself away from the truck, stretched, and walked around to the driver's side. It was getting late. He needed to get moving before darkness made it difficult to navigate the rough two-track road leading back to the ranch house.

When he arrived at the parking area, Denny pulled over and stopped. He'd be returning tomorrow at an early hour and wanted to make certain he could get through the front gate without problems. The party seemed to be winding down, fewer vehicles were present than had been there earlier in the day, but he still felt uncomfortable about knocking on the door. He didn't want to be viewed as an uninvited guest, a party crasher. Resolve suddenly hit him. *Aw shoot*, he thought; *I've got a job to do*. Exiting his truck he hurried toward the front door. He'd find Larry, negotiate arrangements to enter the ranch in the morning, then head directly to his hotel for some serious number crunching. He hadn't gone far, however, when he spied Emilio walking Dirk's horse, Betsy, towards the barn. Abruptly he changed course. Denny felt far more comfortable talking to Emilio than to Larry. He also knew the ranch hand would be an early riser and was more likely to be of assistance in the morning.

The two men greeted one another and Denny explained his concerns.

"Tell you what," Emilio offered. By now they'd reached the paddock and he removed Betsy's saddle, setting it onto a fence rail. "I may not be around when you get here. Lily and I plan to head down to Abiquiu before sunrise. When you get to the gatehouse there's a keypad next to the phone. The access code is 1-2-3-4. When the gate's open just drive on through; it'll close automatically once you're in."

"Thanks," Denny said with relief, "and don't worry, I won't blab the security code to anyone."

"Huh, it wouldn't make no difference if you did or not. The code's the default that came with the gate system. Lots of people know that; them gates are pretty common in ranches around here. Hey, I saw you up on San Cristobal Hill today fiddlin' with your surveying gear. How'd it go?"

"Just fine," Denny replied. "I still can't get over how beautiful this place is. In this weather it was a real joy to be out there, so quiet and peaceful."

Emilio nodded his assent, "I'm a blessed man. Ain't a day goes by I don't think about how lucky I am to work here doin' what I do. This place is like home to me. 'Tell the truth, I don't like that Dirk is selling part of it, but there's nothing I can do about that."

"Speaking of the land sale, I was afraid I might meet your irate neighbor, Mr. Benavides, up on that hill, but he never showed up—thank goodness."

"Actually, it don't surprise me he wasn't around. In fact, you were lucky to meet him th'other day. Most days he's in Arenas Rojas at the nursing home with his wife. 'Bout three years ago Maria had a stroke and although Augusto tried to take care of her he just couldn't manage on his own, what with runnin' the ranch and all. I know they got money problems. Augusto told me not long ago that he's worried about losin' the ranch." He gave Betsy a pat on the neck and began to lead her to her stall. Denny politely offered to carry the saddle and blanket to the tack room.

When the horse and gear were put away, Emilio said, "Come on over to the house with me and I'll introduce you to Dirk. He's leaving for the set tomorrow and probably won't be back for at least a week. You should meet the fellow you're actually workin' for while you've got the chance." Denny fell into step and the two of them made their way back towards the parking lot. As they walked, Emilio continued to fill Denny in on events of the past.

"After Augusto's wife Maria had her stroke, we were all pretty shook up. She's a fine woman: laid back, sweet, one of the nicest folks you could ever meet. She loves company and was constantly inviting us over to visit. Lily, Bella and I would ride all around both ranches and end up over at the Benavides place eatin' fresh baked bread or muffins, and she made the world's best coffee in an old percolator she'd heat on their wood stove. Yes sir, those were good times. She and Augusto are about as close as a couple can be, and Augusto knows he's a lucky fellow to have married a gal like that. She's still sharp as a tack, keeps the staff at the retirement home on their toes, but the stroke left her weak on her right side. It was bad enough back when it happened that she required a wheelchair to get around, needed help getting' dressed, such like that. I've heard she's doin' better now though. She likes to walk around the facility with a cane, and Augusto told me a few weeks ago he's gonna bring her home soon. Another problem with the retirement home is that they only have a few employees who speak good Spanish. Maria grew up on that ranch and most people don't realize how isolated it is, or at least was. For hundreds of years the folks who lived around here spoke nothing but Spanish, up until the 1940s or so. Maria's an old timer, the last of a generation who speak the ancient Castilian Spanish of the conquistadores and who used to rely mainly on barter and trade to survive. She and a handful of others represent a vanishing way of life. Even Augusto is more of what I call a crossover type; he grew up in Las Vegas, New Mexico, spoiled by indoor plumbing, speakin' both Spanish and English."

Denny recalled from his morning research at the County Clerk's Office that the Benavides property had been in the name of Alfredo Peralta until the mid-1970s. "Maria's maiden name wouldn't be Peralta by any chance?" he asked. "A quit claim deed was filed transferring ownership to Augusto and Maria Benavides about twenty years before part of it was sold to Dirk Stetson."

"Yep, Maria Peralta. She had a sister, Conrad's mother, who died young, so the ranch was passed on to Maria and her husband Augusto. That was long before I started working here, of course. All that don't seem to have anything to do with Dirk sellin' part of his land though."

"No," Denny agreed, "it shouldn't. I'm curious though, what does Maria think about the sale? Is she as upset about it as

Augusto? Does she think that part of the Stetson ranch really belongs to her?"

Oh, yeah. She's as pricked about it as he is, but she don't vent her frustrations the way he does. She's a religious sort, keeps burnin' her devotional candles hopin' God will set things right. What can I say? All of us has different ways of dealin' with situations." He motioned Denny towards the front door. "Let's go see what's up with old Dirk."

Emilio led the way through the huge great room amid smiles and nods of greeting from the revelers. Near the back, just outside the dining area, stood a tall man in his 60s with slicked-back dirty blonde hair streaked with grey along the temples. Even Denny, ignorant as he was about Hollywood, recognized the rugged countenance with its strong clefted chin, high cheekbones, and blue eyes. Anyone old enough to remember the old Marlboro cigarette ads couldn't forget Dirk Stetson.

Denny also recognized the person talking to Dirk. Their eyes met and Larry immediately broke away from Dirk and moved across the room, swaying and spilling his drink en route.

"Why, hello, Emilio, it's an unexpected pleasure to see you and Mr. Carswell here at our little soiree. Is everything okay back at the horse barn?" He paused to sip his drink while awaiting Emilio's reply.

"Everything's fine, Larry. We just dropped in so I could introduce Denny here to Mr. Stetson."

"Oh, yes, why that's quite appropriate. So, is Lily minding the store then? I'm sure there's lots to do prior to your drive to the set tomorrow."

"Larry, you know Lily's got a class tonight. And don't worry about the horses, they're fine."

"Oh, if you say so that's good enough for me. I'm just trying to keep tabs on things, making sure my employees are putting in a full day's work, not sloughing off."

Emilio's tone turned sour, but he remained composed, "Mr. Cooper, I can assure you Lily does more than her share of the work around here and spends at least forty hours a week doin' it. And I've told you before, the fact she's getting an education is something she should be admired for. Folks around here should be supportive of her." Pausing for effect he continued in a low, slow voice, almost a growl, "Now there's one thing I would never tolerate in an employee and that's drinkin' on the job. Anyone drunk ever started messin' with my stock I'd kick his ass, if you get my meaning."

"There's no reason to get testy, Emilio. I've got a ranch to run and I'm just trying to keep up with things." He turned in a huff and stumbled, intoxicated, towards the kitchen.

"Sorry you had to hear that," Emilio said. "Ranch politics is all it is. My mother told me to always look for what's good in people, and Lord knows I try, but so far I ain't seen nothin' good in that little turd. He wouldn't bother me none if he didn't have such a cobb up his butt about Lily."

It wasn't any of Denny's business and he knew it, but curiosity forced him to ask: "What did Lily do to upset him?"

"Beats the hell outta me. One day she asked Dirk something about a bill we got for supplies and feed and the next thing she knew Larry jumped all over her ass. He said she should go to *him* with questions about ranch operations; it was *his* job and she shouldn't bother Dirk with details like that. Ever since that day he's been ridin' her like a cheap whore. What gets my dander up, 'sides the fact that Lily's a good kid, is that she's had a hard life. That woman has a lot of good in her and she deserves a break." He took a deep breath to calm down.

"Hey, Dirk," Emilio motioned for the actor to join them. "This here's Denny Carswell. He's the surveyor we got to handle the land deal."

Following Emilio's introduction, the three of them made small talk. Dirk seemed genuinely interested in what Denny thought of the ranch.

"You've got a small piece of heaven here, Mr. Stetson. Emilio showed me around yesterday and I was very impressed. It's a gorgeous spread, and that horse of yours, Betsy, is a real gem."

"Ha, yeah, I know what you mean. Old Betsy's a feisty one, a fine spirited animal. You got to keep your feet in the stirrups if you don't want to fall off her, that's for sure!" Denny nodded slowly as Dirk rambled on animatedly. He seemed pleased to have someone other than his movie peers to talk to. "As you may know, I'm an actor, have been for darn near fifty years. I started when I was just a little shaver 'bout so high." He held his hand at waist level to emphasize his point. "And the toughest scenes to film are those when I'm riding. Keeping my balance on a horse, while trying to remember my lines, requires real concentration. Wendy, one of my directors, calls it 'multi-tasking.' You ever have to do any 'multi-tasking' in your job Mr . . . er . . .?"

"Carswell," Denny said. "Please call me Denny."

"And you be sure to call me Dirk."

"Will do," said Denny, "and, about the multitasking, uh, not that I can think of," he added modestly. "I can empathize with your acting problems, though. I'm certainly no thespian by any stretch of the imagination, but my one experience in a high school play introduced me to the demands of your profession."

Dirk replied with a sly smile, "I know what you mean. The movie business is loaded with thespians. Hell, Rosie O'Donnell is a good friend of mine and I had no idea she liked women 'til she announced it to the world. Man, if that wasn't a kick in the pants. Nowadays most of them are what you call 'out of the closet,' but sometimes you never know."

A tall, well-endowed blonde woman walked up to Dirk and grasped his elbow, pulling him aside. "Dirk, baby, I'm bored," she whined. "You promised to show me around your ranch."

"Not now, Vernie," he glanced at Denny and winked, "I got business to tend to—you know, guy stuff. Why don't you go get us a couple 'o beers." He patted her on the rump, "I'll give you the grand tour in a few minutes." She pouted and gave him a displeased look before striding towards the kitchen, the clicking of her four-inch heels echoing in the cavernous room.

Denny happened to notice Bella sitting by the hearth, a vantage point from which she had a commanding view of the entire gathering. She was with the bald-headed man he'd seen talking to her the night before. They had been chatting, but now they were both intently watching Dirk's female friend with looks devoid of animation. Denny got the impression there might be a certain "lack of harmony" in the Stetson household. He turned to his host, "It sure was nice to meet you Dirk. I've got to go and process today's survey data, but I do want to ask you something before I leave. I'd very much like to know how much of your ranch you intend to sell. Larry's response was somewhat vague when I asked him about it yesterday."

Dirk knitted his brow and contemplated the question. Denny swore he heard gears grinding as the man's thoughts progressed. Emilio was right. His boss seemed rather clueless. "That's a good question, um, Mr . . . ."

"Denny."

"Yeah, Denny. Well, uh, Larry's been bird-dogging this as you know. I usually take his advice on ranch matters, but," he hesitated again, "I don't think we want to sell more than a few thousand square acres—more or less."

"Gee, Dirk," Denny recognized the need for a diplomatic response, "things seem to be a bit 'unclear' at the moment. Maybe the appropriate way to proceed is for you, me and a tribal representative to walk through the western end of the ranch and set markers on the ground where the mutually agreeable boundary will be. By the end of the day tomorrow I'll have recorded the precise coordinates of each existing property corner. When the new monuments defining the land that's for sale have been set, I'll reference them on the same datum, draw up a plat—that's the official diagram describing the property division—and the rest will be up to the lawyers."

The vacant look on Dirk's face indicated a need for further diplomacy. His buxom friend was returning with two over-filled glasses of beer, shuffling along in a desperate attempt to keep from spilling the contents. Foamy liquid splashed onto the beautiful pine flooring with each step, and Denny concluded that a fast exit would be wise. "Of course you might be busy on the set tomorrow and it's too much to ask that you stay around to manage the survey. I'm sure Larry can handle the details with me. Thanks for your help, Dirk, and good luck with your movie."

THE DARK FORM PADDED noiselessly amongst the trees. Slinking towards the doorway it stayed just out of sight, pacing nervously back and forth, lupine eyes glowing yellow in the moonlight. Inside the rustic structure a solitary individual sat beside an old potbellied stove, warming his hands near the surface while he waited for the tea kettle on top to whistle.

The cabin hadn't been inhabited for years until the owner, desperate for help on his ranch—help he was unable to pay for—had allowed the strange young man to stay there, rent free, in exchange for the performance of certain chores. Although the young fellow was of little help tending the grazing stock, he was fit and strong, and could mend fence, chop wood and perform strenuous work that the old man could no longer handle. Thus was born the relationship between Xeno and Augusto Benavides. It was, for the most part, a harmonious alliance. Sometimes the chores would lag when Xeno mysteriously disappeared for days at a time, but he would always return and make up for his absence. He was cheerful and laid-back, and Augusto found himself relying more and more on Xeno so that he could spend more time with his ailing wife. It was, nonetheless, a tenuous friendship. Xeno's odd ways: the silly foil-lined hat he wore, his talk of communication with creatures from other worlds, and the supply of electronic junk that seemed to breed in every nook and cranny of the old cabin, were disconcerting to Augusto. All things considered, however, he appreciated having the young man around.

On a chilly October night such as this, the stove could easily replenish heat that was lost through gaps between the logs, holes created as old cement chinking crumbled and fell away. It could not, however, compensate for the open doorway. Xeno shivered as he waited for his nocturnal visitor.

"Hello, Henry," he muttered quietly as the coyote finally gathered the nerve to poke his head into the room. He never crossed the threshold. That was a respected boundary, set by Xeno having urinated on the doorjamb, a message well understood by his canine friend.

From his pocket Xeno produced a piece of beef jerky and held it up for view. He lifted his upper lip to expose his front teeth and said, “Smile for me, Henry.” The coyote mimicked the gesture, raising his lip to expose long, sharp incisors, a snarl without malice, a manufactured “smile.”

“Good boy,” Xeno acknowledged. He tossed the treat towards the door. Henry caught it in mid-air, then lowered his head and stood silently, waiting for another morsel. Thrice more the procedure was repeated: smile, smile, treat. When he deemed the ceremony complete, Xeno stood and Henry sprinted away into the gloom of night. This odd ritual between man and wild beast had its origins in a bitter cold evening the previous winter. Xeno had spotted the emaciated coyote pup lying beside his truck, shivering. Warmth from the engine had apparently lured him into normally dangerous proximity to human habitat. The animal was in bad shape. If he didn’t intervene, Xeno knew his gaunt visitor wouldn’t make it through the winter. He began feeding the patient scraps of food. Night after night the animal came by and Xeno had dutifully fed him, nursing him back to health. But he gradually decreased the amount of food, forcing upon his charge increasing levels of self-reliance. Now the visits were less frequent. They’d devolved into short training sessions. The handouts were small—tasty rewards for Henry’s one trick, a toothy grin.

DENNY KNEW HE HAD a full day’s work cut out for him as he slowly drove the primitive road back to the paraje on San Cristobal Hill. It was early morning and the Sun had just risen above the horizon. As he passed the stables he could see lights on in the adjacent living quarters, and both of the large trucks used to ferry horse trailers back and forth to movie sets were still parked where they had been the evening before. Apparently, Emilio and Lily’s departure had been delayed.

As he again assembled the components of his GPS base receiver over the corner, Denny went over in his mind his tasks for the day. Once the base station established good satellite links, he would carry another receiver, tethered to an antenna atop a long pole, to other corners and reference points around the property. This secondary receiver, called the 'rover,' collected signals from satellites orbiting overhead as well as signals via radio-link from the base station. This arrangement would supply Denny with coordinates of the other locations, relative to those of the base station, to within one hundredth of a foot—three millimeters—in near real time. The so-called 'real time kinematic,' or RTK, technology he was using had recently revolutionized the discipline of surveying. In many cases it obviated the need for angle and distance measuring devices entirely. It made this job much easier, but the system wasn't without limitations. The eastern part of the ranch, including two corners along the border with National Forest land, lay in heavily timbered, rugged locations. He doubted his old rover's receiver would be able to establish satellite link at those two corners. Indeed, later that morning, after an enjoyable hike throughout the wooded portion of the property while carrying the rover, his suspicions were confirmed.

Undaunted, and knowing that the ranch survey was 80% complete and slightly ahead of schedule, Denny drove back to the parking lot. Grabbing his total station, tripod and retro-reflecting prisms, he prepared to finish the survey the old fashioned way. He had to traverse to the remote corners, those that had been inaccessible by GPS, turning angles and shooting distances until he tied in to one or more locales for which he had GPS coordinates. The work would be strenuous, the terrain rugged, and he was faced with running line for almost a mile. But Denny was in a great mood. The weather was perfect and the scenery magnificent. These were the times Denny truly appreciated his job; this is what he lived for: hiking in gorgeous country, making high precision measurements—and getting paid to do it!

He flagged a nail and set a temporary point near the ranch house as a point of beginning. He'd start by taking 'side shots' to the near corners of the ranch house, barn, paddock and the living quarters used by Emilio and Lily. These data would be needed in order to locate these permanent structures on the plat. From there he'd commence his open traverse to the outlying corners, eventually working his way over to points on his GPS baseline. While in the process of setting a prism by the barn, he noticed Emilio walking towards him. "I expected you to be half way to Abiquiu by now," said Denny as they shook hands. "What happened?"

"Andre Bivens, the director of the film, looked me up to tell me there were problems on the set. They won't be ready for the horses for at least another day, maybe even two or three."

"That's too bad," Denny said as he finished leaning the prism pole against one of the corral posts.

"Oh, it don't matter to me. Andre was pissed though, says the weather is gonna break for the worse in a couple days. He's worried they may have to postpone some of the scenes with people on horseback for a week or more. I guess that's gonna cause some financial headaches." He watched as Denny walked back to the total station, made a couple of quick measurements and returned to relocate the prism. "Say, Denny, I've got free time on my hands, you want some help runnin' that prism?"

Denny smiled, "I'd love some." He briefly described what he was doing and the two men set to work. With Denny at the gun and Emilio shuttling the prism around according to Denny's instructions, they made quick work of acquiring data for the ranch structures. Emilio offered to help with the ensuing traverse, but Denny diplomatically refused. He knew that part of the job would go almost as fast without a novice assistant. Besides, he'd learned enough about ranching to realize Emilio was being overly generous with his time; there were numerous chores to be done around the stables.

The landscape through which the traverse was to be run was more rugged than Denny remembered from his brief foray of the previous day. He had to set many more points than he'd anticipated, leap-frogging his total station and tripod with two prisms and their associated bi-pods. Short sight lines through and around regions of dense scrub oak and small fir trees hampered him, but progress was steady and his spirits were high. In short,

Denny Carswell was having a blast. Once again he reveled in working under a cloudless, blue late October sky in some of the prettiest country on Earth. Occasionally he'd catch a break when he could run a segment many hundreds of feet through an old horse trail or along an open ridge.

The first corner monument he sought turned out to be a rusting iron pipe covered with vestiges of yellow paint. Without knowing precisely where to look he might not have found it. Situated deep in a gully and obscured by a tangle of oak, he wondered why anyone would have selected such a spot for a property corner. A moments thought, however, made him realize whoever did the work had no choice. It was on an extension of the nearly due north-south line that separated Taos and Mora counties. In addition, it served to mark a corner between USFS land and the private parcels owned by Stetson and Benavides. Fortunately Denny didn't have to occupy the point with the total station. He hacked out a line of sight with his machete, set the prism directly on top of the pipe and shot his angle and distance from the ridge above the wash.

Most of early afternoon had been spent getting to that first remote boundary marker. He now had to find the other one that, according to his forest service TOPO map and Josh Marquez's plat, was located on a steep slope that faced to the southeast. He thus had to hike through heavy forest and up and over a steep ridge to get to it. The effort of carrying his equipment uphill felt good. His muscles had warmed up nicely, and as he made his way to higher ground, the intense Sun made him perspire. He stopped briefly to remove his light jacket and tie it around his waist. He took a long drink from his canteen and noted that, if it weren't for the fall colors, the golden aspen and yellow oak leaves, he would swear it was mid-August and not late October.

It was late afternoon by the time he made it to the last corner. Denny carefully recorded the pertinent measurements in his field book. He always hand-entered important data even though all of the angles and distances he obtained using the total station were also stored in instrument memory. In general, the electronic data were those he used to perform his office calculations and to prepare plots. He could effortlessly download everything to his PC and process the numbers using any of several good software packages. But Denny was also a purist. He prided himself on the neatness and clarity with which he recorded his field notes, and on several occasions, when he'd encountered instrument problems, the handwritten data had saved him from repeating days' worth of labor.

He trudged back up and over the hill with the tripod on one shoulder and the instrument case in the other hand, stopping occasionally to catch his breath. *One more set up*, he muttered to himself. He'd planned the traverse so that the last shot would be from the open knoll he'd seen from the paraje San Cristobal when he and Emilio had toured the ranch. Last evening, after dismantling the GPS base antenna, Denny had replaced it with a retro-reflecting prism which he'd aimed at the knoll. This final measurement would tie the traverse into the paraje.

Denny felt enormous relief as he contemplated today's accomplishments. The survey was almost finished. Sure, he would close the traverse later on, probably in the morning given the lateness of the hour. But that was just a check: a long sight from his point of beginning at the parking lot up to the paraje. Tonight he would have all the data he required to fix every corner of the property onto a precise geodetic datum. Any monuments he set to partition the land according to the wishes of Larry and tribal officials could then be accurately and precisely located with respect to the rest of Dirk's ranch.

On top of the knoll Denny set a temporary point and hurriedly set up his tripod. It was almost five PM and daylight was limited. Something he hadn't anticipated was that this last shot would be made late in the day looking to the southwest. The Sun was just off line but its glare was going to make viewing difficult. As he took his backsight and then rotated the total station's telescope in the direction of San Cristobal hill, he heard a sharp crack, the unmistakable sound of a gunshot. At the paraje, more than 2000 feet away, he could just make out the outline of a figure standing near his prism. He looked through the scope, squinting against brilliant sunlight slashing obliquely through the side of the objective lens as he brought the scene into focus.

Despite the disruption of the image by scattered light, Denny could clearly see two figures. One was on the ground with his hand in the air. The other stood over him holding a rifle. Suddenly he saw a flash come from the end of the barrel followed seconds later by another bang. Yet a third shot was fired and Denny instinctively recoiled in horror as the fellow on the ground fell backwards and lay still. When he again looked through the eyepiece, the gunman was walking rapidly away. All Denny could glean about the assailant, as the man stole away into the setting sun, was that he wore a wide-brimmed hat and a long, heavy coat. Looking back at the other individual lying next to the paraje, he noticed a shock of white hair. Although he'd been unable to identify the assassin, he had no problem recognizing the victim. Denny Carswell had just witnessed the murder of Augusto Benavides.

VII

Chaos reigned on the Stetson ranch once word of the murder spread. Denny had feverishly attempted to call the ranch house on his cell phone, but, when there was no response, he'd immediately called 911. He'd then run to the top of Cerro Medio as fast as his 65-year-old legs and the rugged landscape would allow. It was Augusto all right. Two clean bullet holes in his head had ended his life only a few feet from Denny's prism. When he realized he could be of no assistance to Augusto, Denny had walked all the way back to the parking area in fading daylight. He arrived just after a sheriff's vehicle and an ambulance had gotten the attention of the ranch inhabitants.

Now, an hour later, there was quite an assembly of spectators at the paraje. Denny's truck had lead the way up the rugged service road to the top of the hill. A sheriff's deputy followed close behind. Over the next hour, everyone who had been at the ranch and received the shocking news had made his way to the site. Emilio, Lily and Bella arrived on horseback while the others: Larry, Dirk, Dirk's friend Laverne, Conrad Sussman and a heavyset woman who worked part time as a cook, trickled in, either on foot or by car, bearing shocked looks on their faces. Everyone knew the victim, and by now they all knew he'd been gunned-down in cold blood.

The deputy did his best to cordon off the scene and maintain order. He was a young, wiry man named Paul who was one of two Taos County sheriff's representatives assigned to the town of Arenas Rojas. He reminded Denny a little bit of Barney Fife, at least when it came to looks. The hawk-like face and seemingly oversized gun belt held up by bony hips were reminiscent of the comedic, overzealous deputy sheriff of Mayberry, NC. All likeness stopped there, however. Paul was a consummate professional in the way he handled the enormous responsibility that had suddenly been thrust upon him. He expertly controlled the area around Augusto's body, allowing no one to get within fifty feet. His strict demeanor left no doubt that he was in charge, and Denny soberly suspected that 'deputy Paul's' service revolver contained more than the token one bullet allotted to his erstwhile look-alike 'Deputy Fife.'

BOBBY WADE, sheriff of Taos County, New Mexico, was sitting at his desk immersed in paperwork when he received word of the homicide outside of Arenas Rojas. "Son of a bitch," he cursed after talking to the 911 dispatcher. It was 5:30. Dinner time, and he'd promised his wife he'd come home at a reasonable hour so they could spend "quality time" together. That's what happens when your spouse takes one of those touchy-feely classes in sociology or psychology or whatever the hell it was at the community college. Those ivory-towered knuckleheads were polluting Carmen's mind with all kinds of sappy 'relationship' bullshit, something that he, the County Sheriff, was powerless to prevent. "Hell," he sighed as he stood and grabbed his hat and gun belt. His wife would have to eat alone tonight. She was impressionable, that was true enough, but she was also reasonable, and he didn't expect her to give him a hard time once she learned the reason for his absence. Murder is the highest crime there is, and he'd be derelict in his duty as the county's chief law enforcement officer if he failed to respond properly to this call. The sheriff was a responsible man.

He strode to his vehicle and pulled out of the parking lot, heading towards the two-lane highway that would take him to the Stetson ranch. Along the way he contemplated the nature and timing of this latest crime. There was another murder, yet unsolved, that was consuming nearly all of his waking hours. Three weeks earlier, a young boy had disappeared from the nearby Taos Pueblo. Days after a massive search effort was launched, the child's body had been found by fishermen deep in the Rio Grande gorge: the cause of death—knife wounds to the back and chest. The brutal nature of the crime, the youth of the victim, and the fact that he was Native American had brought considerable attention from the press. The tribe was exerting ever more pressure on his office as time passed and no leads developed. That pressure was nothing, however, compared to the pressure Bobby Wade placed upon himself. He was a man who believed in action and swift justice, especially when it came to crimes against children. The crushing frustration he felt regarding that crime had put him in a persistent foul mood.

Now his life would be further complicated by another murder. At least the killing of an old rancher wouldn't attract the same sort of attention from the press. With any luck this one would have clear motive and obvious suspects. Bobby considered himself to be a clever problem solver, someone who had a sixth sense when it came to the workings of the criminal mind. Now was the time to put that ability to work, to solve this latest iniquity so that he could continue his other investigation. The situation awaiting him at the Stetson ranch carried with it one positive factor: the presence of Emilio Sisneros. Emilio and Bobby went way back, all the way to their late teens when they worked as ranch hands right out of high school. For a while they'd both dreamed of hitting it big on the rodeo circuit, driving overnight to compete in nearby weekend events and pushing to get back to their day jobs before they got fired. When reality finally set in, they'd gone separate ways, Bobby signing up with the sheriff's department and Emilio continuing to do what he loved most: riding horses and roping steers. Now much more sedate in their mid-fifties, the two rarely saw one another. This year their annual hunting trip had been cancelled because of Emilio's responsibilities on that movie set and because of the heinous murder of that Indian boy. Bobby took comfort in knowing that if Emilio knew anything about this killing, his judgment and insight would be of great help.

Sheriff Wade soon made his way up the hillside. The drive from Taos was not insignificant, but the sheriff hadn't broken any speed records getting there. He obviously had faith that his employee, Deputy Paul Viera, could handle the situation until he arrived. Upon Wade's arrival, the two lawmen conversed briefly before the sheriff took charge and turned to address the assemblage of onlookers. A man of medium build, Bobby Wade had the rugged countenance and sinewy constitution of someone who was no stranger to hard work, and he exuded a no-nonsense air of absolute authority.

"Folks, what's happened here is serious. I want no one to touch anything here or go anywhere near the body. And nobody and I mean NOBODY, leaves until he's given a statement to either me or Paul. Is that clear?" His eyes penetrated to the very soul of everyone there. No one was going to disobey this man. He paused to spit and then pointed to the tripod with the prism on top. "Which one of you is the surveyor?"

Denny stepped forward with his hand half raised. "That's me, sheriff. My name is Denny Carswell. I'm the one who called 911."

The sheriff motioned him aside. "What else can you tell me about what happened here?"

"I was witness to the killing, sheriff. When I heard the first gunshot I looked through my instrument, the one set up over on that far ridge." He pointed towards the knoll, now completely lost in darkness. He shrugged, "I immediately called for help and then ran to see if I could help Augusto."

"So, you saw the killer?"

"Yes, Sir."

"Well, shit. Save me some trouble here," he said with exasperation. "Who was it?"

"The thing is, sheriff, I couldn't identify him. All I can tell you is that he wore a heavy coat and a wide-brimmed hat."

"But, you could tell who the victim was, right?"

Denny explained about the angle of the Sun and the poor relative placement of the killer. The sheriff gave Denny a hard stare. Then, admonishing him to 'stay right here,' he walked over to consult with his deputy. The two men made no effort to keep from being overheard and all in attendance listened intently to what they had to say. Paul revealed what he'd learned while he'd awaited his boss's arrival.

"Like I said earlier, sheriff, I feel Mr. Carswell, the surveyor, is the key. He was the only person in the vicinity and he actually saw the crime take place." He waited for the sheriff to speak, and continued when no questions were forthcoming. "The rest of these folks were all present on ranch property, but they were way down where the buildings are. That fellow there," he pointed at Dirk, "he's Dirk Stetson. I'm sure you've heard of him—famous movie star. Anyway, he owns the ranch. At the time of the killing he claims he was rehearsing dialog for the movie he's making." He pointed at Laverne, "She's his personal assistant and acting coach—says they were together the whole time and didn't hear anything."

Paul pointed to Bella next. She wore a stylish denim shirt and slacks that were inadequate to counter the evening chill. The temperature had dropped by more than ten degrees since the Sun went down, and a steady breeze was now blowing across the exposed hilltop. She hugged her shoulders against the cold.

"This is Mrs. Stetson," Paul said. There was no need to say more; he was quite certain his boss knew who she was.

"How'd do, ma'am," the sheriff said with a tip of his hat and a slight smile. His hard-boiled attitude eluded him briefly as he addressed the beautiful woman. "Would you be kind enough to tell me where you were during all of this and what you know about it?"

Bella bit her lip and shook her head slowly. Tears welled in her eyes and her chin trembled when she spoke. "I just can't believe Augusto's dead," she began with a sob. "He was a good neighbor and a good friend . . . ." She paused to compose herself. "And poor Maria, she's so sweet and . . . ." She sniffed and took a deep cleansing breath, letting it out with a shudder. "At the time the deputy arrived I was just getting ready to go inside. I'd been out on the patio, sunbathing. The weather's been so nice you know. When Larry opened the door and I saw it was the police, I got dressed immediately and came up here to see what happened and to see if I could be of any help."

"You say you were outdoors sunbathing. Did you happen to see or hear anything—like gunshots?"

She thought for a moment and shook her head. "Not that I can recall, but I wasn't really paying attention. Gunshots are a common sound around here as you probably know; people do a lot of hunting and target shooting in the National Forest."

The sheriff nodded and looked over at Paul. His assistant recognized his cue and continued his briefing. He pointed again. "This fellow here is Lawrence Cooper. He's what I guess you'd call the ranch administrator. Bella and Dirk are gone much of the time and he stays around and runs the place."

Illuminated by the harsh light from the headlights of Sheriff Wade's car, Larry looked deathly pale. He was staring off into the distance in an intoxicated stupor. At the mention of his name he turned and looked vacantly at the sheriff.

"Mr. Cooper, you're the one who my deputy met when he got here 'bout an hour ago, that right?"

"What? Oh, I guess so. I mean, yes, it was me."

The sheriff frowned, "Tell me, Mr. Cooper, did you start drinking before or after you shot Mr. Benavides?"

"Huh?" Larry's expression slowly turned from stoned to incredulous, "Sheriff, surely you can't believe that I . . . of course I didn't kill anyone. How can you even suggest that?"

"Just a shot in the dark, so to speak. You'd be surprised how many criminals will confess to their actions when they are cognitively impaired. Maybe you had something to do with this and maybe you didn't, but as ranch manager that puts you under suspicion." He raised his voice, "For the record, I consider everyone here to be a suspect, with the exception, perhaps, of my friend Emilio here." He looked at Emilio and asked, "Anything you can tell me about this?"

"Hell, Bobby, I don't know any more'n you do. People here have known Augusto for years. It's hard for me to imagine anyone at this ranch killin' him." He shrugged, "All I can think of is that his death may have somethin' to do with a land deal. Dirk wants to sell part of his ranch to the Indians in Santa Teresa. Augusto was pretty irate about it. Me and Denny had a run-in with him about it yesterday. Augusto thinks . . . er, thought, that he's the rightful owner of the part Dirk's gonna sell."

"You see or hear anything, Emilio? I mean at the time of the killing."

"Lily and I were workin' in the barn gettin' tack and supplies together for our trip to the movie set. Filming was postponed for a day, but things are back on schedule." He paused to think, "We were talkin' while we worked, and of course we were inside makin' some noise. I don't recall hearing nothin'."

Lily nodded her assent, confirming Emilio's version of events. Sheriff Wade asked if anyone had anything else to add, and was about to dismiss the group, when a shout came from Paul. The deputy had begun to walk around the area with his flashlight, checking the crime scene perimeter and looking for clues. Something had apparently caught his attention. All eyes were on the young man as he stooped to retrieve a large object from behind a boulder. When he approached the sheriff all could see that he was holding a rifle by the tip of its barrel.

Paul beamed with pride as he leaned the weapon against the side of his patrol car. He knelt and examined it in better light. "Looks like we have our murder weapon, boss, an old .22 caliber Remington 572 pump. Nice gun. And what's more," he could barely contain his exuberance as he stood, "the serial number is intact."

A queasy feeling came over Denny when he saw the rifle. Abruptly he broke from the throng and walked quickly to his truck. Several others, including Sheriff Wade, watched him go. After rummaging around in back of the driver's seat he returned on wobbly legs, his face white as a ghost. He said simply, "It's my gun, sheriff. The killer must have taken it from my truck while I was working on the survey."

Gasps came from many in attendance at Denny's pronouncement. Sheriff Wade's expression hardened and an uneasy quiet descended upon the paraje. One could hear the faint rustling of aspens in the distance punctuated by an occasional click as the engine block of the sheriff's patrol car cooled down. After reflecting for a moment on Denny's bombshell, Wade pushed himself away from the fender of his car and surveyed the people around him.

"The M-E is on his way from Taos to retrieve the body. He'll have a forensics team with him. I want everyone to go on back to the Stetson house and remain there until I'm satisfied with the statements we've recorded." He turned to his deputy, "Paul, put the gun in the trunk of my car and bag it carefully so's not to smudge any prints. Then I want you to tend to Mr. Carswell." Bobby Wade reached into a pocket of his utility belt and pulled out a set of handcuffs. "Mr. Carswell, your truck has to stay here 'til the forensics guys have gone over it. You, sir, have the right to remain silent, and a lawyer will be provided to you if you so desire."

Sheriff Wade continued to Mirandize Denny while he cuffed the surveyor's hands behind his back. Denny was stupefied. He was like a deer caught in headlights. The sheriff's speech and muffled exclamations from others blended as unintelligible noise while his brain attempted to cope with what was being done to him. He stumbled slightly, weak from blood rushing from his head to his core—he was in shock. Recognizing the plight of his prisoner, Sheriff Wade helped Denny into the back seat of his patrol car and allowed him to rest for a moment in the open doorway.

"Bobby," the sound of Emilio's voice made it through the thick haze surrounding Denny's thoughts. "Are you sure about this? Denny here is about the last person I'd suspect as a murderer. He was just out here doin' his job."

The sheriff listened to his friend before tipping his hat back. "I was thinkin' that too, amigo, but in my business you can't ignore facts. You said yourself that it's hard to imagine anyone else at this ranch doin' any harm to his neighbor. The way I see it is anyone could have killed Augusto at any time, but the crime didn't happen until this feller showed up. Besides, he was the only one who was out here this afternoon; everyone else was down in the ranch buildings, and they all have alibis. He says he saw the killing, but, very conveniently, he could identify the victim but not his assailant. That seems darn near impossible in my experience. I suspect he's just tryin' to deflect suspicion from himself and onto the mysterious man in—how did he put it?—a hat and a heavy coat. But the most significant facts, Emilio, are that it was his gun that was used in the murder, and that the two of them had an argument yesterday. You yourself were there."

Emilio thought for a moment. "It still don't make any sense to me, Bobby. That argument wasn't personal. Old Augusto wasn't mad at us, he was just upset about the land sale."

"And how do you know exactly what happened way out here on this hill late this afternoon? 'Could be those two had a knock-down-drag-out when Carswell was settin' up his equipment." He motioned in the direction of the prism and tripod. "You know as well as I do, Emilio, how easily an argument can get outa' hand, and if Carswell's gun turns out to be the murder weapon, I wouldn't be doin' my job if I didn't arrest him."

Emilio sighed in resignation. He looked briefly at Denny and then gazed downward. There was defeat in his voice. "I know the facts, Bobby, but, hell, my gut tells me you're wrong, and, besides, everyone at this ranch knows how to load and fire a gun. I still say he's tellin' the truth."

"I agree with Emilio." It was Bella who spoke up, and her tone was adamant. Everyone looked at her in surprise. "Mr. Carswell is no killer. I'm a good judge of character, sheriff, and he's not the type to do something like that. Even if he had a volatile temper, I can't think of what would motivate him to kill that poor old man."

"Well, ma'am, you and Emilio may be right. But let me remind you that it's my job to investigate crimes that are committed in Taos county and to arrest those who are, or appear to be, responsible. It's up to the DA to decide if the State will prosecute this man, and ultimately, if things go to trial, a judge and jury will decide on both his guilt and his fate."

She bit her lip and nodded. "Okay, I guess you have to act on the evidence at hand, circumstantial as it may be." Looking over at Conrad, who had been as quiet as a church mouse since he'd arrived atop San Cristobal hill, she said, "I'm going to make sure Mr. Carswell obtains proper legal help. Conrad, I want you to represent Denny."

For the first time that evening Conrad Sussman's face registered something other than disinterested boredom. He arched his eyebrows and inflated his chest. "I *am* rather busy, Bella, but (sigh) I suppose as a favor to you I can serve as Mr. Carswell's attorney." He addressed Denny, "I can make time to meet with you in the morning. I'll have a number of questions to ask you so that I can better prepare a defense. Depending on the court schedule we could have you out on bail by as early as tomorrow afternoon. No promises, of course, I can't control what the judge will say."

"Tomorrow afternoon?" Denny fumed, "Does this mean I'll have to spend the night in jail?"

"I'm afraid so, Mr. Carswell. That's the way the system works. The wheels of justice turn slowly, sir, but they do turn . . . eventually. Don't worry. I'm good at what I do, and I'll do a good job as your advocate. By the way, I charge $300 an hour plus expenses. I just want to clear that up before we enter into contract."

"Oh, for Pete's sake, Conrad, just do your job," Bella interjected. "Bill all charges to Dirk and me. I feel personally responsible for Denny's predicament. If he's innocent, and I'm sure he is, I don't want him to suffer any more indignity than he already has."

The sheriff put an end to further discussion and directed everyone to leave for the ranch house. Flashing lights were visible in the parking lot; the medical examiner and the forensics team had arrived. Sheriff Wade sent Emilio down to guide the ambulance back up the trail to the paraje. He then ordered Paul to stay with Denny in the cruiser out in the parking area while he finished questioning the others. In the back of his mind Bobby Wade was calculating how long it would take to finish this damn business and get home to his wife. They might yet get to spend some quality time together this evening. He was even beginning to believe that would be a good idea. Maybe there was something to this relationship crap after all.

VIII

The ride to the Taos County jail was perhaps the worst experience of Denny's life. With his hands manacled behind his back it was difficult for him to sit normally. His physical discomfort, however, represented only a small part of the agony he was enduring. Denny had been in a police car only once before, and that had been as a passenger, not a prisoner. The back of the cruiser was a mini prison; a heavy grill separated him from Sheriff Wade, who was driving, and the doors were locked to prevent egress—the same feature that kept toddlers from accidentally opening doors and falling out. *They treat us murder suspects like children*, he thought dolefully as the car sped through winding, hilly country in one of the blackest nights he could remember. Sheriff Wade had nothing to say, and they rode along in silence for what seemed to Denny an eternity. Finally he couldn't take any more.

"Sheriff, I have very expensive equipment set up in the field around the Stetson ranch. I also have valuable gear in my truck. Not only have you arrested an innocent man, you're jeopardizing my livelihood."

"I'm gonna caution you one more time, Mr. Carswell. It'd be best for you if you don't say anything without your lawyer present. As far as your surveying gear goes, it'll be kept safe and sound as evidence. The forensics guys, my 'clue crew' as I call 'em, will take good care of everything."

Exasperated, Denny suddenly lost his appetite for conversation. Quiet was better than talking to a hardheaded idiot. He began to contemplate his immediate future. From cop shows he'd seen on TV he knew he'd be permitted one phone call. The question now was, who to contact? He desperately wanted to talk to his wife, Pamela, but she was attending a conference out of town.

All he would accomplish by calling her would be to upset her. Moreover, she couldn't be of any real help outside of rendering moral support. He'd already made arrangements to meet with Conrad in the morning, so phoning him would be redundant. Trying hard to relax, Denny leaned against the car door. Soon the lights of Taos came into view, and within several merciful minutes he was out of the car being escorted into the police station.

The indignity of being arrested and handcuffed was nothing compared to the insult to Denny's self-esteem as he endured the "booking" process. *What fun*, he thought as he was forced to stand next to a yardstick to be photographed and then put through the awkward routine of being fingerprinted while still in handcuffs. The sheriff stood by and inspected the fingerprints carefully before nodding his approval and drifting away. Denny overheard him tell a subordinate to make sure copies were made available to the "clue crew" first thing in the morning. *Boy, it sure would be damning if they found my fingerprints on my truck, my equipment or my gun*, he thought ironically. He felt like shouting to the sheriff to make sure to get reference prints from others so that they might find real evidence to act upon should they run across the real killer's prints. He restrained himself when he realized, despondently, that challenging the sheriff's methods and ego would almost certainly do more harm than good.

A uniformed woman whose nametag read "Baldonado" handed him a clean orange jumpsuit and instructed him to change in a small restroom near the holding area. She then handed him a canvas bag with a tag on it. "Place your personal clothing and valuables in here, sir. We'll keep them under lock and key until you get out."

He mumbled something resembling gratitude before shuffling into the restroom. He reemerged moments later looking like a death camp survivor. The jumpsuit, size XXL, hung around him like a flag on a calm day. The short sleeves of the garment extended well beyond his elbows and he'd had to roll up the leg cuffs so that he wouldn't trip over them. The figure who confronted Ms.

Baldonado and handed her his bag of personal effects looked so comical and pitiful she couldn't keep from laughing.

"Sorry about the uniform," she said. "Until the laundry comes back from the cleaners this will have to do. Don't worry, though, we should have something in your size in a day or two."

Her smile was disarming. For the first time in hours, Denny relaxed. He must indeed look pretty outlandish. If only he'd been booked and photographed after he'd been forced to change into this "tarp," the picture would have been a great memento to give to Pam. Officer Baldonado had even called his prison garb a "uniform," one she would be able to replace in *only a day or two*. The absurd idea that he'd be walking around for that long, looking the way he did, put Denny over the top. He started laughing, and, as he thought more and more about his ridiculous predicament, he laughed harder and harder. Denny laughed like he'd never laughed before. His infectious behavior soon elicited a Pavlovian response from those around him. Everyone in the booking area was now smiling and giggling, looking at the strange old man draped in an orange tent who seemed to be having the time of his life. When he finally settled down, Denny felt one hundred percent better. A good laugh does a body good, he realized. Feeling almost human again, he decided it was time for introductions.

"Ms. Baldonado, my name is Denny." He held out his hand. She hesitated briefly before shaking it. Experience and training had probably cautioned her against informal social contact with prisoners, but instinct ultimately prevailed as she submitted to Denny's friendly overture.

"Hi, Denny, I'm Camille Baldonado. You don't seem . . . ." She was flustered. This man was accused of murder, but he really wasn't what she would have expected a killer to be like. "I hope things work out for you."

He nodded soberly.

"Well," Camille drew a deep breath and exhaled sharply, "I guess we'd best get on with things." She beckoned to a fat, elderly guard to approach them. "This is Wayne. He'll escort you through the wing where the male prisoners are held. Good luck, Denny," she added as he was led away.

DENNY COULDN'T SLEEP. The Taos County jail was a beehive of activity at all hours of the day. Voices and noises drifted through the corridor that led to his cell. He was bone-tired and emotionally drained from the day's events, but the noise and strange surroundings kept him from getting any real rest.

At just past one AM another prisoner was brought in and placed in the adjacent cell. There were three cells in his "block" and two were now occupied. Denny refused to contemplate what would happen if four or more people required incarceration. He didn't fancy the idea of having a roommate.

The new man's name was Roy. He was a large fellow both in height and girth—his "uniform" fit like it was tailor made. Long, stringy brown hair and a full beard conjured up an image of a biker sans colors. He seemed to be enjoying himself, chatting up the guard, Wayne, as he settled into his quarters for the evening.

"Hey, Wayne," Roy shouted after the guard had locked him in, "tell Camille I got something for her if she gets lonely."

Several more explicit statements followed, but Wayne refused to acknowledge any of them. Such were the interactions of prison life, Denny surmised; Wayne's heard it all before. No sooner had the guard left than Roy addressed the only other person within earshot. "Yo, old man, what are you in here for?" Denny remained quiet hoping this latest annoyance would stop yapping if he got no response. He was wrong. The fellow kept babbling, telling Denny how this must be the hundredth time they'd arrested him, rambling on about the various faults of the facility and its staff.

Denny finally reached his limit. He had to do something to get this guy to shut up. "I'd like to get some sleep here, amigo. If I tell you why I'm here will you leave me alone?"

"Yeah, sure, man. So, what despicable act got your ass thrown in here?"

"They've accused me of committing homicide."

"No shit! Hey, are you the dude that whacked that old rancher 'bout twenty miles from here?"

"That's what I'm accused of, yes," Denny replied, "but I didn't kill anyone."

"Right, I hear you. And for what it's worth I didn't split open Jimmy Chung's head with a bottle neither." He chuckled and Denny saw a pair of heavily tattooed arms emerge through the bars of Roy's cell, coming to rest on a cross piece. Roy was standing as close as he could to Denny's cell, preparing to escalate their conversation.

"Oh, God, what have I done?" Denny groaned. His loquacious new acquaintance was just getting warmed up.

"Not that the little shit will press charges or nothin'. Jimmy knows I'd whip his ass if he did that. Plus, I know stuff he'd rather I not talk about, if you get my drift. These walls have ears, man, so let's just say that, hypothetically speaking, Jimmy's livelihood may not be a hundred percent legitimate."

Roy droned on. For the most part, Denny tuned out. There were grandiose references to drug and alcohol use, narrow escapes from the law, and something about trying to collect a very large reward. Eventually all the talk blended into white noise as Denny's eyelids drooped and then closed for good. He fell into a troubled sleep punctuated by a strange dream in which he had become a nocturnal predator, a Dr. Jekyll/Mr. Hyde monster who suffered amnesia in the mornings that followed his bloody escapades. After each villainous deed he would awaken to accusations and blame resulting from a trail of circumstantial evidence pointing to him as the killer. His pleas of innocence were unable to convince any member of society that he wasn't responsible for any of his actions: not law enforcement, his closest friends, not even his wife. All the while, however, deep in his subconscious, he knew he was guilty.

The clanging of a metal door at the end of the hallway woke him up. For a brief moment Denny wondered in panic if, in fact, he really had killed Augusto Benavides. The nightmare had seemed so real that in his current half-awakened state facts mixed with fiction. Soon the cobwebs in his mind dissolved to reveal the absurdity of his thoughts, and Denny got up off his cot, yawned, and walked the few steps to the steel bowl set into one wall that served as his toilet. The smell of fresh coffee was in the air. Reality was once again in charge of his conscious. He was in jail, innocent of the charges against him, and breakfast was about to be served.

A bowl of rubber-like oatmeal, buttered toast, orange juice and coffee, all neatly arranged on a steel tray, were passed to him through a slot in his cell door. *Yumm*, Denny thought as he tried a spoonful of gruel. His cheap plastic spoon broke when he tried to lift the sticky mass from its cheap plastic bowl. *No dangerous weaponry here—these are definitely something the TSA would allow on aircraft*, he mused as ironic assessments of his situation flooded his thoughts. He was in a foul mood. Deciding to forgo cold oatmeal, he chugged his OJ and placed all but the small Styrofoam coffee cup back onto the tray, shoving the trash back into the pass through.

The morning dragged on. It was soon 10:00 AM, then 10:30, 11:00; where the hell was Conrad? Hadn't his lawyer said he'd be by in the morning? Confinement was getting to Denny. There was nowhere to go and nothing to do except sit in this 8 x 12 cell and stare at the walls. It wasn't a place for an innocent man.

He tried to relax and ponder the events that had gotten him here. The vision of Augusto, hand raised in a last second plea for mercy, jumped into mind. Denny closed his eyes and examined the scene. Even as a memory, the Sun seemed to bother him as he shifted to the heavily clothed figure with the gun. How big was the man? How tall? Thinking hard, he was able to place the figure into view alongside the prism pole and tripod. He'd set the prism so that he could sight over it to aim at his total station. That would put the top of the prism maybe 5'6" above the ground. The big question then became: how tall was the murderer relative to the prism? Denny went over and over the scene in his mind, trying to conjure up every pixel of the image. The hat was a problem. It made it hard to determine the killer's height with any precision. Denny was reasonably sure, though, that the killer was about average height, maybe 5'8" or 5'9". If it was someone who was at the ranch last night it could still have been anyone—except Dirk Stetson. Dirk was well over six feet tall and the more he thought about it, the more certain he was of Dirk's innocence. *Well, that's one down and . . . how many to go?* he wondered. He made a mental list: Dirk, his friend Laverne, Conrad, Emilio, Lily, Bella, Larry, and the fat lady who worked in the kitchen. The cook could be ruled out too, Denny thought, especially if she wore a coat. She was much heftier than the figure holding the rifle. This last thought raised another possibility; the murderer could be a woman. Denny hadn't seen any part of the killer's face, and it took no masculine strength to pull a trigger or work the pump action of a gun. If the killer were female, that meant Lily, Bella, and Dirk's bimbo friend, Laverne, were all legitimate suspects. Denny couldn't envision any of them pushed to the point of taking a human life, but he couldn't rule them out with certainty.

That left Larry, Emilio and Conrad. Larry was presumably drunk at the time, and neither he nor Conrad, especially the latter, would have had the stamina or the will to climb Cerro Medio, kill Augusto, and return to the ranch unnoticed. And Emilio? He'd expressed regret regarding the sale of the ranch. Denny's impression was that the ranch was more of a home to Emilio than it was to either Dirk or Bella. Would the thought of losing part of it be enough to send him over the edge, to kill a troublemaking neighbor who posed a threat to his current idyllic job and lifestyle? *No,* Denny reasoned, *Emilio wasn't the type.* Besides, the man had defended him during the questioning session Sheriff Wade had subjected them to up on Cerro Medio. Come to think of it, Bella had also disagreed with the sheriff when he'd cuffed Denny. If either of them was the killer wouldn't he, or she, want suspicion thrown onto someone else? *Hell,* Denny fretted, all of the people at the ranch that evening had legitimate alibis, and none, as far as he knew, had sufficient motive to kill someone.

His musings turned to motive. The land sale was undoubtedly a big factor in all of this, and Augusto had been threatening to stop it. There was a tidy sum of money associated with the sale, and money was a powerful incentive. Denny mentally kicked himself for overlooking the obvious: Santa Teresa Pueblo.

The real money wasn't what Dirk and Bella would be paid for the land, it was the millions the pueblo stood to make once the casino was built. He'd heard the tribe had been trying for years to acquire the bottom land along the highway. The surprise visit from Tommy Hawkfeather and his large assistant, outside of Alice's restaurant, coupled with their kindly offer of assistance, were testament to their extreme interest in the land.

Denny rubbed his eyes and lay back on his cot. He was becoming more and more dejected. If the killer was from Santa Teresa, he felt it would be next to impossible to bring him to justice. Tribal police would become involved because any evidence trail would necessarily involve two jurisdictions, the pueblo and the county. Based upon what he'd seen in Sheriff Wade, his "my way or the highway" personality, he imagined any relationship between Wade's office and another law enforcement agency would be strained if not outright antagonistic.

As Denny lay wallowing in self-pity, his colleague, Roy, began to make noise. The man had just woken up and was in a disagreeable mood.

"Yo, jailer, how 'bout some breakfast over here? You tryin' to starve us?"

A guard came walking down the corridor. He was young and thin, and wore a clean, pressed uniform. "Settle down, Roy. I'll bring you some lunch in a few minutes."

"'An I could really use a smoke. You got any cigarettes?"

"Can't smoke in here, Roy, you know that. It's against the rules."

"Hell, I don't give a shit about rules, man. I smoke. Okay? You may not like it, but it's legal, ain't it? It's my Constitutional right. You gonna deny me my rights given by the US Constitution?"

"Sorry, Roy. Just sit tight."

"Aw, come on, man, just one coffin nail. I won't smoke in here if you don't want. Put the cuffs on me and let's go outside. It's a beautiful, sunny day out there; it'll only take a couple minutes. I get into a bad mood when I don't get my nicotine fix."

"You sure know how to sweet talk me, Roy, but the answer's still 'No'."

The guard left while Roy continued to vent frustration at not being allowed to smoke. Denny ignored him and resumed his analysis of Augusto's murder. He decided to start over, recounting everything that had happened since five PM the day before. A new resolve was setting in. Sheriff Wade wasn't going to solve this case. He had his suspect in custody and was now focused on strengthening the case against Denny rather than searching for, and pursuing, other leads. The Lord helps those who help themselves, he remembered soberly. If his name were to be cleared, he'd have to do his own detective work. Somehow he needed to come up with information to help his case, a difficult proposition while confined to a jail cell with an annoying neighbor.

A puzzling thought surfaced and Denny knitted his brow. Something Roy had said to the jailer moments earlier commanded his attention. He sat up straight, his mind whirring, extrapolating his line of reasoning. Roy was still jabbering, but Denny wasn't listening. *That's it,* he thought excitedly, *that could be the first chink in the killer's armor.*

Lunch arrived. The young guard unceremoniously shoved a box into the pass-through. Denny sighed and retrieved it. Hunger was getting to him. He hadn't eaten anything substantial since yesterday at breakfast; his stomach was empty and it was affecting his thought processes.

Popping the box open, he stared at the contents: a bologna sandwich, a small apple, a large cookie laced with what looked like macadamia nuts, and a carton of apple juice—exactly the mix of items one would expect to find in a first-grader's lunch box. He peeled off the top bread layer of the sandwich to reveal a single slice of bologna topped with mayo and badly wilted lettuce. Sniffing the mix he deemed it fit to eat, but, just to make sure, he pulled out the lettuce and set it aside. It looked more like silage than it did something out of a garden—no use taking chances. He took a bite and wrinkled his nose—not exactly haute cuisine, but as long as he could chew, swallow . . . and forget, his blood sugar would eventually return to its normal range. The apple looked good so he set it aside. He'd relish it later as dessert. Along the Rio Grande, just downstream from where he currently sat, grew some of the best apples he'd ever tasted. The fruit were in season now. He was thankful for this small, crisp reminder of civilized life, a vestige of normalcy that transcended the rest of his prison nightmare.

Denny turned his thoughts back to what Roy had said. He arose and began pacing, analyzing the events of the past two days and organizing them into something that might explain, in large part, what this was all about. There were gaps that needed filled-in, but a plausible scenario was emerging. The killer's identity was still in doubt, and what he had was still only a theory with lots of loose ends, but Denny was growing ever more excited. He tried to calm down, reminding himself that there were still questions that needed answers, answers that were required before he could present his arguments to the sheriff. Once he was out on bail he'd be free to pursue his hunches, but there was, nonetheless,one thing he could do right here in his jail cell. It was important, something that would either lend support to his hypothesis, strengthen it, or else refute it and thereby send him back to square one. Being able to act, to do something to help his cause rather than mope around and grumble, had a tremendous cathartic effect. He couldn't stay calm.

"Guard!" he yelled. He pressed his face against the bars of the cell. Most of the door at the end of the corridor was hidden from view, but he could tell it had been propped open. He shouted again. Someone had to hear him. To his right he saw two tattooed arms emerge from Roy's cell. His outburst had piqued the interest of his neighbor.

The young guard who'd delivered breakfast and lunch made his way down the hall.

"I was wondering if you could do me a huge favor," said Denny as his captor approached. He wanted to stay on this fellow's good side so he was being excessively obsequious. "I really, really need to see my field notes from work I did yesterday. They're in an orange, hard-covered book that's about four and a half inches by six and a half inches. Is there any way I can gain access to it? It may be very important to my case."

The guard frowned. "We have some magazines if you want something to read, but I don't know about that book you want. Where is it?"

"It was in my vest which is, or was, in my truck," Denny replied. "It may now be where you keep your evidence."

The guard's eyebrows elevated, "Evidence, huh? There's no way you're gonna get access to that. How about I bring you an issue of 'Time' magazine?"

"Hey, buddy," Roy spoke up from the adjacent cell, "you got the 'Sports Illustrated' swimsuit issue? If so, I got dibs on that."

Denny was exasperated. "Can you at least ask the sheriff if I can look at that notebook?" When he realized what he was asking for, to examine evidence in a murder case—one in which he was accused—he did a double-take. "Okay, just ask the sheriff if I can get a photocopy of the entries from yesterday. I don't need the original, just the information, and please tell him it may be extremely important." Denny made eye contact in an effort to reinforce his last statement.

A slight nod from the young man was a good sign, and Denny smiled with gratitude. The jailer then turned to Roy. "The judge is in court now. Your mouthpiece is waiting to take you to your arraignment. I suggest you make yourself presentable."

Denny sat down after the guard departed. He felt satisfied he'd made an impression, and, perhaps, had started down a path that would broaden the investigation.

Roy's lawyer, a public defender, soon entered the hallway accompanied by the guard. He was a short, middle-aged man dressed in a custom tailored, pin-striped suit. Threaded through the collar of his shirt was a loud bow tie. Denny chuckled inwardly at the sight of the dapper fellow. He had the look of a stereotypical advocate, replete with a suede briefcase and wingtips, a regular Perry Mason. As the lawyer passed Denny's cell he paused briefly, "Are you Mr. Carswell, the man accused of murder?" Denny was surprised the man was talking to him but nodded in reply. "Your case is to be heard right after ours. Hasn't your attorney shown up yet?"

"No," said Denny.

"Mmm, that's most unfortunate. Look, if he doesn't come by soon, I'll ask the judge to recess for ten minutes so that we can talk. Arraignment is an important part of the criminal judicial process. Considering the crime with which you are charged, you MUST have adequate representation."

Denny thanked him and watched as Roy was handcuffed and led away by the guard, his lawyer a step behind. It was kind of the PD to offer his assistance, and Denny was truly appreciative, but the revelation of his impending date with the judge begged the question: where the hell was Conrad?

Half an hour later Conrad arrived. He was ushered into the cell block by none other than Camille, the woman who had assisted in booking him the night before. Denny smiled at her and immediately, before he even acknowledged Conrad's presence, asked her to contact the sheriff about gaining access to his field notes. He figured that if enough people bugged the sheriff about that, the more likely he was to get it.

Out of respect for attorney-client privilege, Camille withdrew, leaving Conrad standing just outside the cell door. He seemed anxious and agitated and greeted Denny with a perfunctory: "We have a problem."

The comments from the public defender had left Denny understandably disturbed and troubled. He vented his anger at Conrad's apparently flippant attitude towards his predicament. "Where the hell have you been? If you knew my arraignment was at 2:00 PM, why weren't you here an hour ago?"

"Hey," came the equally vehement reply, "you're lucky I'm here at all. Remember, I'm a busy man. I'm only representing you as a favor to Bella; she wanted you to have the best solicitor in Taos, and believe me," he sneered for effect, "I'm the best."

Denny almost suggested that Conrad perform a physically impossible auto-erotic act, but decided this was the wrong time to antagonize the one person who, at least in theory, was working on his behalf. Besides, he didn't know anything about judicial protocol; maybe Conrad was right. "All right, I'm sorry," he sighed, "So, what exactly do you mean by, 'we have a problem'?"

Conrad pulled a bottle of nasal spray from his coat pocket and took a hit in each nostril. The poor guy sounded like he was coming down with a head cold. "I spoke with the district attorney when I came in. It seems he is, er . . . was, a good friend of Augusto's—thinks the guy was some kind of fucking Saint. Anyway, he's taken a special interest in this case. He's gonna try his best to get the judge to deny bail."

The air rushed out of Denny's lungs like he'd been gut-punched. This was unbelievable. Things were going from bad to worse. He couldn't possibly spend another day in jail. Confined to this cell he'd be unable to work towards clearing his name. "You've *got* to get me out of here, Conrad. All of the evidence against me is circumstantial, you know that, and I have no prior criminal record." He sounded desperate. "I'm also innocent," he added emphatically. "People like me should never be subjected to such treatment in our society." A door slammed and he turned to see Camille making her way down the corridor. It was time for his first appearance before the judge.

IX

The courtroom reminded Denny of a church. Dark brown carpeting deadened their footsteps as he and Conrad walked to the front and sat directly behind Roy and the PD. In the back of the room, on either side of the door they'd come in, were two captain's chairs, one of which was occupied by Sheriff Wade. The sheriff looked bored and refused to make eye contact as Denny went by. National and State flags flanked the judge's bench and a large clock hung on the wall directly behind him. Its prominent placement was to remind all in attendance, except perhaps the judge, who faced the other way, to be punctual, direct, and succinct while conducting court business.

At the moment, the honorable Benjamin Hinojosa was perusing documents relating to Roy's case. He was a small, round-faced man, perhaps sixty years old, with thin, receding white hair. The large robe draped over his shoulders made him look even smaller, almost child-sized. Denny felt a kindred attachment to the judge in a fashion sense, although his bright orange jumpsuit made a much bolder statement than did flat black. Hinojosa's eyes moved rapidly across the pages. As he read, the frown on his face deepened.

"Mr. Hazelton, sir, you are a menace to society. This list of infractions is, to put it bluntly, both impressive and intolerable. For most of your thirty-two years you've been awfully busy breaking the laws of this great State, and for reasons I dare not contemplate, your latest 'victim' has declined to press charges." The lecture continued for several minutes. Although Judge Hinojosa focused primarily on Roy's latest infraction, the assault for which he was currently charged, he also cited specific past incidents—those that illustrated a disturbing pattern of behavior. In the end, Roy was charged with disturbing the peace and, on advice of counsel, pleaded guilty. Given the choice of further incarceration or a fine he could not afford, Roy was marched back to the jailhouse to remain a guest of the State of New Mexico for several more days. The judge then turned to Conrad and Denny and motioned for them to approach the bench.

"Dennis Carswell, you have been arrested on suspicion of murdering Augusto Benavides. How do you plead?"

Conrad spoke up, "We plead not guilty, your honor."

Hinojosa removed his reading glasses and addressed Denny. "Sir, your presence here today before my court is not to determine your guilt or innocence; it is to determine whether or not there is probable cause to detain you."

Denny repeated his plea as if Conrad hadn't spoken, "I'm not guilty, your honor."

The judge nodded, then turned to the district attorney, "On what evidence has this man been arrested and imprisoned?"

The DA referred to a yellow legal pad as he replied. Denny had expected an older man commensurate with the knowledge that the prosecutor had been a friend of Augusto's. This fellow was middle aged. He seemed fit, although not particularly athletic, had bushy dark-brown hair, and a prominent nose. The suit he wore was clean and pressed but old and somewhat faded. His mouth was surrounded by an impressive Hulk Hogan-style mustache, and when he spoke he projected his voice as would a stage actor or the host of a game show. His words were carefully chosen, delivered with eloquence and perfect diction. "Your Honor, Mr. Carswell's detention is based upon evidence obtained within an hour of when the murder occurred. He was the only person known to have been present on the hilltop where the crime took place. He had had a heated argument with the victim the day before, and preliminary forensic analysis indicates that his gun was the murder weapon. In addition, he claims to have witnessed the murder and gotten a good look at the victim, but, surprisingly, even though the killer and victim were only ten feet apart, he was unable to provide any descriptive information about the killer." The DA then paused and gave Denny a cold, hard look. Turning his attention back to the judge he spoke more slowly, continuing his presentation with added emphasis and passion. "I have in my hand, your honor, a copy of the medical examiner's autopsy report on Augusto Benavides." He began to read verbatim from the document. 'Cause of death: two gunshot wounds to the head. Powder residue fixes the distance from the weapon at six to ten feet. The lethal shots followed an earlier gunshot wound to the left side of the upper torso . . . .'"

The judge looked at Conrad during the reading of the report, but the lawyer seemed oblivious to what was being said. Instead of listening he was waving at something, perhaps a gnat, which was pestering him. Hinojosa interrupted the DA. "Mr. Figueroa, this is not a trial. I just want facts to support your arguments for probable cause."

"I know that, your honor. The reason I'm quoting the autopsy report is perhaps premature, but I'm trying to promote the heinous nature of this crime in order to strengthen the State's request to deny bail. It was an execution, sir, pure and simple, three gunshots delivered at close range, the final two constituting a coup-de-grace. That, coupled with the previously stated evidence, should convince the court of the danger this man poses to our community. We feel that in light of the evidence bail should be denied."

Denny spoke up in his own defense, earning him stern looks from both Hinojosa and Sussman, "Your Honor, please, all of the evidence is obviously circumstantial. I did witness the murder, through my transit scope from more than two thousand feet away. The killer wore a wide-brimmed hat and a heavy coat. The facts as presented by the district attorney are skewed because . . . ."

The judge banged his gavel on the bench. "That's quite enough, Mr. Carswell." He looked at Conrad. "Counsel, what have you to say about the evidence presented by the State?"

Conrad waved away another bug before responding, "Like my client said, your honor, everything that links him to the crime is circumstantial. I can't stress that enough."

There was a pause as the judge continued to stare condescendingly at Conrad. He asked, "Is there anything else you'd like to say on your client's behalf—regarding bail perhaps?"

Conrad merely repeated himself, "Well, like I said, your honor, my client pleads not guilty, and the evidence is entirely circumstantial. Bail should not be denied."

"I see." Judge Hinojosa sighed and addressed Denny, "Although the facts linking you to this crime are indeed circumstantial, I do believe there was probable cause to detain you as a suspect. Furthermore, the callous, pre-meditated nature of the murder suggests that, if in fact you are guilty, you should remain incarcerated until your preliminary hearing. Bail is therefore denied."

Denny was incredulous. "But, you honor, I'm the one who reported the murder to the police. I ran over to try to help Augusto after I witnessed the shooting. Sir, I'm not capable of such murderous behavior; if you check police department records you'll find that I've never been accused of any type of crime save for one speeding ticket, and that was almost thirty years ago! And, your honor, with all due respect to Sheriff Wade, he's convinced I'm guilty. I'm afraid further inquiry on his part will only focus on incriminating me instead of pursuing the real killer. If I remain in jail, I'll be unable to conduct any sort of investigation to try to clear my name. The motive for the murder lies behind a land deal, your honor. I just know it, and as a surveyor I . . . ."

The judge cut him off. "Sir, the one thing I don't want to happen is for you to interfere with Sheriff Wade. I've known Bobby Wade for years. He's a smart and dedicated law enforcement officer, and I trust that he'll conduct a thorough and impartial investigation." Denny began to protest, but the judge again silenced him with a sharp rap of his gavel and an icy stare. "Mr. Carswell, I've made my ruling. Now, as your attorney has undoubtedly told you, this is nothing more than an arraignment. Within the next ten days you will be summoned to a preliminary hearing at which you will either be indicted, that is formally charged in writing, of this crime, or you will be released. If an indictment is returned, then you and your counsel must either enter into a plea bargain or the case goes to trial." Hinojosa placed both hands on his bench and surveyed the room, meeting the gaze of each player in this evolving high-stakes game of criminal murder prosecution. "It is also my duty to inform the accused that New Mexico is a state in which first degree murder can carry with it the death penalty. Our collective duty, gentlemen, in light of the grave consequences of our endeavor, is to . . . get . . . things . . . right!" The veins in the judge's neck stood out and his last three words were uttered in a low growl. "One life has been taken and another hangs in the balance." He slammed his gavel down one last time and glowered, "You are all dismissed."

The death penalty. In ten days he would undergo his preliminary hearing, a process referred to in some states as a 'Grand Jury'. His indictment was almost assured, and the DA was going to push for the death penalty in the subsequent trial. Denny was hopelessly scared. He now sat in a folding chair in an anteroom of the courthouse, his face buried in his hands. Conrad sat across from him, having just delivered the bad news of the district attorney's aggressive prosecutorial intentions. The preliminary hearing would be conducted here in Taos County, but the trial, an event that would almost assuredly take place, would be held at the district courthouse in Santa Fe. *Great,* Denny thought, *I'll be tried for murder right in my own back yard. My wife, friends, and neighbors will all be present when I'm sentenced to die by lethal injection.*

Two guards arrived who would take him back to his jail cell. For the next ten days he'd remain a prisoner in the Taos County jail. After that he would be transferred to the State Penitentiary outside of Santa Fe. The shackles around his feet clinked as he walked, the sound mingling with the footsteps of the guards echoing through the short hallway that led to the building's back entrance. As they emerged into the bright sunlight of another gorgeous fall day, Denny's spirits picked up a bit. Something about being outdoors in the crisp autumn air with fallen leaves blowing underfoot and a cloudless blue sky above, injected him with a modicum of hope and resolve. In a flash of what he recognized as self-preservation, he concluded his only hope for survival was to take matters into his own hands if at all possible.

Conrad had followed him out of the courthouse and was now walking towards the front parking lot where he'd parked the Lamborghini. Denny asked the guards to wait and yelled to his lawyer, "Hey, Conrad, we need to talk. Can you meet with me back at the jail? It's important."

Conrad shrugged but nodded in assent before walking away. Denny watched him briefly and frowned. The "first appearance", his arraignment, had gone poorly. He hadn't been particularly impressed with Conrad's performance or lack of preparation for what he now realized had been an important step in the judicial process. A good lawyer would have pushed harder for some sort of bail arrangement, he reasoned, but Conrad's heart wasn't in this job. And, there was something else about his behavior that troubled Denny; the lawyer's actions and mannerisms just weren't right in some way. Camille, the friendly guard at the sheriff's office, had mentioned to him how fortunate he was to be represented by Conrad. She was of the impression that the man was well known and respected in legal circles. Denny had his doubts.

Soon after they returned to the jail, Conrad was allowed to see him. Wayne, the larger of the two guards with whom Denny was acquainted, actually let Conrad into the cell after frisking him. Apparently Denny wasn't deemed to be a risk to his own lawyer. Conrad took a seat on the cot while Denny stood and paced. He was agitated and nervous and needed to work off some energy. His mind was also working in overdrive. Since being reminded of what might be a looming death sentence, his survival instincts were keeping him incredibly alert. His thoughts returned to the scenario he'd come up with prior to his arraignment. It was time to conduct his first interview. The request to meet with Conrad wasn't related to his legal defense. He was, instead, betting that his lawyer might be a source of valuable information.

"Thanks for standing by me during that arraignment, Conrad. Some of the folks here at the jail tell me you're pretty good. I just want you to know that, well, my attitude earlier today was bad because I was scared and upset. What I'm trying to say is, 'I'm sorry'."

The lawyer held up his hands and smiled, "I understand, no hard feelings on my end. Your case is truly circumstantial. As soon as we're through here, I'll bring Bella up-to-date on what happened at your first appearance. Even though bail was denied, I think things went pretty well. I'm going to suggest she put me on a $50K retainer for the work I'll be doing to prevent indictment at your preliminary hearing."

"Good, good, that sounds great," said Denny, forcing a smile. "Say, Conrad, I'd like to talk a bit about Augusto and the land sale if that's okay. Maybe we can come up with some leads as to who might have wanted him dead. It would be a real feather in your cap not just to win an acquittal for me but to help solve the crime as well. You'd be a regular Perry Mason! Of course, I would also be forever in your debt if you could pull that off."

Conrad broke into a sly smile. His pupils were dilated and he was rapidly tapping his foot on the floor. "That's the plan," he said in a stuffy voice. "We'll have no trouble shooting holes into their arguments. Like I said, their case is purely circumstantial, pure bullshit." He sniffed and wiped his nose on the back of his hand, then reached into the pocket of his coat to retrieve his bottle of nasal spray.

Denny waited until he had Conrad's attention and continued, "The other day when I first met Augusto, he was disappointed that you were acting as Dirk's counsel in the land deal with the pueblo. He implied you two were related. Is that true?"

"Yeah, but not by blood. He was my uncle, my mother's sister's husband."

Denny nodded thoughtfully, "What can you tell me about him?"

Conrad shrugged, "We didn't interact much. I've always been really close to my aunt Maria though. She and my mom were not just sisters, they were best friends. Even after we moved to Santa Fe, when I was a small child, my mother would bring me to Maria and Augusto's ranch at least once a month. As a kid I wasn't into horseback riding, hunting, fishing, you know, outdoor shit. But Aunt Maria always made me feel welcome just the same. She was a great cook and she'd always give me some kind of present for coming to visit . . . *heh, heh,* even though I had trouble understanding what she said."

"You don't speak Spanish?"

"Not much. My dad didn't speak it so he forbade my mother from teaching me. He didn't want us conversing around him if he didn't know what was being said. At the time I thought he was being a real asshole, but now I can kind of see his point: no use bothering to learn something you'll never use. Anyway, my dad's partner, the guy he started his law firm with, Tom Poole, he speaks good Spanish, so the rest of us partners don't need to."

Son-of a-gun, Denny suddenly thought, *I should have realized,* "So, you work for that firm that has all those billboards around Albuquerque, 'Sussman and Poole'?"

"Correct," Conrad smiled, "We're the largest accident consulting firm in the State. Our signs and commercials really rake in the business. Would you believe our profits more than *tripled* when we got into late-night television sponsorship? By the way, the ads we're running now were my idea. You might have seen them, the ones with the hot chick, showing just the right amount of cleavage, begging for legal help with her mangled car sitting in the background. Dad didn't like them at first, said it was too obvious, you know—blatant pandering to people's greed. But he really changed his tune when the calls started pouring in. Yes, sir," he laughed, "billable hours. That's what it's all about in the tort world, billable hours."

Conrad was on a roll, talking about what was obviously a favorite subject. "And that sweet thing who works our TV spots is truly a sweet thing, if you get my drift. I met her one night in April at that strip club way out east on Central Avenue, the 'Coyote Den.' She was bored with dancing, said what she really wanted to do was be an actress. That's when it hit me, the idea of her hosting our business ads. Well, not to belabor the point, she was ecstatic when I hired her and (heh, heh) ever since she's been extremely grateful."

Still kicking himself for his failure not to associate Conrad's name with those awful advertisements, Denny realized that Camille's complementary statement about Conrad had nothing to do with his competence. It all had to do with notoriety. The man wasn't a criminal defense lawyer at all; he was a damned ambulance chaser!

Denny pictured the mantra on billboards that littered Interstates 25 and 40 in and around Albuquerque.

'WHEN YOUR IN A WRECK DON'T BE A FOOL, PICK UP THE PHONE AND DIAL SUSSMAN AND POOLE.'

The grammatically incorrect use of the word 'your' somehow seemed fitting accompaniment both to Conrad's abilities as his advocate and to the nature of his firm's business. Suddenly he was sick of Conrad's gloating. "As fascinating as all of this is, Conrad, we're a bit short on time. Can you tell me anything about your dealings with the Pueblo over this land sale? Did anyone ever mention Augusto as being a problem? Was he an annoyance to them?"

The lawyer thought for a moment. His mind was slowly breaking away from the pleasant image of his stripper/spokesperson and returning to his client's predicament. "No, not that I can recall. Things were moving along swiftly and smoothly until Uncle Augusto's death. His name never came up in conversation."

Had your Aunt Maria expressed any concerns about his safety? To your knowledge did Augusto ever say anything to her about the land sale—what he was trying to do to stop it?"

"Hell, I haven't talked to her since she first went into that nursing home. That was over three years ago. Huh, thinking back, it's a good thing I went to see her then because that's when I first met Dirk and Bella. When the Pueblo later asked to buy their land, the Stetsons called me and hired me to help them out."

"Didn't it bother you that your Aunt and Uncle had problems with the deal? After all, they are your family."

Conrad became defensive, "Hey, just 'cause we're related doesn't make them right. I looked at those survey reports just like you did, 'Mr. Surveyor'. The land that's in question is owned by

Dirk. My old fart uncle had his say in court and lost. My opinion was, and still is, that the matter was settled. Augusto should have accepted it. You know–life moves on." He sniffed and brought his hand up to his nose to massage his sinuses. He looked uncomfortable.

Denny was now certain that Conrad's suffering wasn't due to a head cold or allergies. He also felt that questioning his lawyer would be of no further use. All he'd learned, really, was that Conrad hadn't visited his aunt after her husband's brutal murder. His lack of compassion seemed strange in light of his professed close relationship with her, language barrier notwithstanding.

It was close to six PM. Receding sunlight and long shadows poked through the west-facing windows of the cells. A police cruiser rumbled by as part of the night shift took to the streets of Taos. Silence fell upon the occupants of the cell. Conrad was focused on something on the far wall, restlessly drumming a beat on his thighs and murmuring words to a song in his head. Denny contemplated Conrad's role in recent events–pushing forward with a real estate deal opposed by his relatives and representing an accused murderer whom he didn't seem to care about. 'Billable hours' and schlock TV ads were all that seemed to motivate this shallow man. He called for the guard and both he and Conrad stood as the latter prepared to leave.

"I'm hungry," Conrad announced with a yawn. "Working late like this always stimulates my appetite. I'll let you know if anything comes up. The judge has to schedule your hearing within ten days."

Wayne arrived and let Conrad out. "Oh, Conrad," Denny said as his barrister began to leave.

"Yeah?"

"You're fired."

"Huh?"

"Get lost. I don't want you spending any more time on my case."

The two exchanged cold looks before Conrad spoke, "A man in your position shouldn't . . . ."

Denny interrupted, "I'll take my chances without you."

The lawyer shrugged and adopted an apathetic look—as though someone had just told him his socks didn't match. He abruptly turned on his heel and left. Dinner was waiting. Denny settled back onto his cot to cogitate his decision to dismiss Conrad. Within the next day or two he'd need a new advocate, and he'd need a good one; the exigency of the preliminary hearing demanded it.

Wayne soon returned carrying two dinner trays, one each for Denny and Roy. He unceremoniously slid one through the access slot to Denny's cell. Denny was hungry. Pulling the tray towards him he surveyed the contents: meatloaf covered in a thin mushroom-based gravy, succotash, and a bowl filled with a creamy, grey liquid—'mystery soup'. The meatloaf was still partially frozen and the vegetables were freezer-burned and desiccated. Denny's hunger disappeared. Underneath his dinner plate, however, he noticed several folded sheets of paper. Wayne stood by as Denny spread them out. They were photocopies of his field notes from the day before.

"Complements of the Sheriff," Wayne said before shuffling away to deliver Roy's meal.

Roy had little difficulty expressing his 'gratitude.' "Yo, Wayne, what gives, man? Looks like somebody took a dump on this plate. You tryin' to kill me and my buddy Denny with this crap? That's another one of my Constitutional rights you've gone and violated: the one that's supposed to prevent cruel and unusual punishment. Hey, Denny, what do you say we go on a hunger strike, starting right now? We'll get skinny and sick in a few days. When the press gets wind of it we'll be able to say 'Thank You', in spades, to these Nazi cops. Even better, give me the name of that shyster you just fired. I bet he could tear 'em a new one, what with all the publicity we'll get. Might even make us some quick cash. What do you say, Denny? C'mon man, you got to help me here; I'm tryin' to think outside the box."

Denny wasn't listening; he was too excited. "Wayne!"

The guard's selective hearing was in good form. Ignoring Roy's rant he looked at Denny, "Yeah?"

"Thanks a bunch for getting me these forms, and thank the sheriff for me, will you? Oh, wait, there's one more thing I need. Can you bring me a calculator, one that has scientific keys? I need to crunch some numbers."

The guard shook his head, "No way. We don't allow prisoners to have stuff like that, for security and safety reasons."

Denny didn't see how a calculator could serve as a deadly weapon, but he knew it was pointless to argue. "Okay, okay, no calculators. Can you bring me a set of log tables instead?"

Wayne frowned, "Sir, that's ridiculous. No extra furniture is allowed in your cell; you've got a cot and a commode, that's all you need. This is a jail, not the Ritz."

"Wayne, you don't understand, I need a set of tables of logarithms, preferably with a table of common logs of trig functions: sines and cosines." The guard's face was blank and uncomprehending. "Well, uh, how 'bout a slide rule?" Denny's voice was low and hollow. A minute earlier he'd been elated to receive his notes, but that jubilation was now crushed by the frustration of not having the computational tools to derive the results he so desperately needed. The emotional cycles he'd experienced in the last twenty-four hours were taking their toll. He reluctantly backed off. "That's okay, Wayne. Thanks again for the notes." He plopped down onto his cot and rubbed his eyes. Taking a deep breath he realized he needed to clean up, take a shower, shave. Picking at his dinner, despair consumed him. Sooner or later he was going to have to call his wife, Pam, and tell her the bad news. She'd find a lawyer somewhere and they'd start preparing a defense strategy. With the police barking up the wrong tree it was going to be an ugly protracted struggle to clear his name, and if the real killer was never brought to justice, a cloud of suspicion would hang over him for the rest of his life, even if, by some miracle, he were to be found not guilty. He buried his face in his hands and massaged his temples. Images of what his future might hold came and went, none of them particularly appealing. A sudden thought flashed through his mind and he fought to capture it. The police. Denny slid his tray onto the floor and stepped up to the cell door.

"Wayne. Hey, Wayne."

The guard appeared moments later carrying the book he'd been reading.

"Wayne, I'd like to make a phone call."

"Can't do that now, sir. Only the sheriff can approve requests for calls, and he only allows them during visiting hours." He started to leave but Denny wouldn't have it.

"I believe I was permitted a phone call when I was arrested, but I never made one. Please remind the sheriff of that; I know he's a stickler for protocol. Let's see what he says."

Wayne sighed and sauntered off. He'd be able to get back to his novel soon enough.

It wasn't long before Bobby Wade came striding into the jail wing. It was 6:45 and Wayne had bugged him just as he was preparing to go home. His wife would be pissed if he came back after 7 PM. She'd called him at lunch to tell him she'd picked up two juicy T-bones at the store and had a big batch of posole slow-cooking in the crock pot, both favorites of the sheriff.

Dinner was all Bobby had thought about since the governor of Taos Pueblo had left his office. The man had barged into the station a few minutes after 4:00 accompanied by two other tribal big shots and a tribal cop. They'd demanded an audience and had proceeded to ream him about the murdered boy. The pueblo was getting more and more stirred up by the day. Wade's problem was: what more could he tell them? The investigation had stalled, and there were two officers, himself and a detective from the town police department, working on it full time. He had also mentioned the murder of Augusto Benavides in the hope it would help him politically. He advised his unexpected guests that his office was now investigating two murders and that his resources were now stretched wafer-thin. He, in turn, was reminded that a suspect was in custody in the Benavides case, and only slightly veiled allegations were made that the murder of an Indian boy apparently didn't command the same attention given to the slaying of a white rancher. Not much could shake Bobby Wade. The physical punishment he'd endured in his rodeo days and the long, difficult hours he'd put in as a ranch hand had hardened him to the majority of life's vicissitudes, but the psychological strain of the inability to solve this boy's murder was becoming almost too much. Now, this surveyor was riding his ass about making a god-damned phone call.

"What's all this bullshit about a phone call?" Wade was all business. He stood outside Denny's cell, hands on hips and breathing hard from the short but quick walk from his office.

Denny heard the strain in the sheriff's voice and sensed the fire in his demeanor. Something was bothering the man, but Denny was just as angry and fed up with his circumstances as Bobby Wade was with his. He was innocent, damn it, and he was tired of being bullied by both the sheriff and the legal system. His eyes met those of his captor and returned the hard stare.

"I've got a right to a phone call, sheriff, and you know it. I want to make it now."

"Oh, you want to make it now, huh? What am I, your servant?"

"Look," Denny fought to keep his cool, "I'm an innocent man accused of murder and you're not doing squat to discover the truth about what happened. It's my right to call someone to help me out. Believe me, Sheriff Wade, I'm going to find a new lawyer and I promise I'll make it as tough on you as I possibly can if you don't obey the law and let me make that call."

Tattooed arms emerged from the adjacent cell as Roy weighed in with his two cents. "Yeah, sheriff, let him make his call. It's his right. Says so in the Constitution."

"Aw shut up Roy!" The sheriff looked at Denny and considered his options. He hated to give in to a prisoner's demands; it threatened his authority. But if he let him make one lousy phone call it would sure make things easy on him, especially in the conjugal relations department. His wife wouldn't fuss too much if he only ran a few minutes late. "Okay, but make it quick. I'll give you five minutes. Put out your hands." Denny stuck his hands through the cell door and the sheriff cuffed him. "Phone's down at the guard station."

As they marched down the corridor Denny said, “I’ll need to look up the phone number.”

Sheriff Wade grabbed Denny’s sleeve and spun him around, “You mean you don’t know the number of the person you’re callin’? Hell,” he fumed, “I don’t run no directory service here. The deal is: no number, no phone call.”

“It’s not going to be a problem, sheriff,” Denny said in a calm voice, “I suspect you have the number readily available.”

X

Lieutenant Horace Begay of the New Mexico State Police studied the memo, biting his lip in concentration. Karen, the officer on duty this evening in dispatch, had taken the message and called him immediately. In his eighteen years on the force he'd never run across anything so bizarre. Now, back at his desk in the barracks south of Albuquerque, he was contemplating what he needed to do. A friend of his was in trouble up north in Taos. Karen had carefully jotted down what the fellow wanted Horace to do, but there was more to the situation than what she'd been told. The sixth sense that some cops develop over time wouldn't let him accept things at face value. Denny Carswell had asked him to get in touch with a longtime mutual friend, a retired surveyor named Bonifacio Romero. Boniface could apparently help him by reviewing survey documents related to a real estate transaction. Almost as an aside, the request came with a terse background story involving Denny being falsely accused of murder. Well, if this were just a big misunderstanding, why hadn't Denny called Boniface directly and left the State Police out of it? No, there was an important reason why Denny had dialed the State Police and, in particular, the substation where Horace was based; his friend was in real trouble.

Having made the requested phone call to Bonifacio, Begay stood and moved toward the door with the fluidity of a dancer. He was a large, thick-chested man who possessed the high cheekbones and flat nose so prominent among the Native Americans from northwestern New Mexico. An engaging smile coupled with an even temperament helped to put those around him at ease, a result that was enormously advantageous in his line of work. His lumbering, avuncular image was, however, entirely misleading. As any number of miscreants could attest, Horace had a sharp eye, a nearly photographic memory, and an astounding agility that would make an NFL scout drool. There was urgency in his step as he grabbed his hat and stepped out into the cool night air. He'd be in Taos before midnight. A look of both concern and anticipation crossed his face as he started the police cruiser. Tomorrow was going to be an eventful day.

ON TUESDAY MORNING a few minutes after 9:00 AM, a small crowd gathered in the booking area of the Taos County jail. Horace Begay had arrived earlier than the others and was now meeting privately with Sheriff Wade in a corner of the room. The two continued to talk in low voices while the other three visitors gave the officers some privacy and conferred amongst themselves.

His work with the State Police had taken Begay all over New Mexico. As he'd risen through the ranks, he'd spent more and more time dealing with other law enforcement agencies. Horace was proud of the diplomatic skills he'd honed in such encounters. Being a good cop was all about people skills. It necessitated being a good communicator, not just when dealing with the public, but also while interacting with one's peers. Horace's political abilities were being put to the test.

As Begay saw it, the problem with many sheriffs' departments was a lack of training. State Police officers attended a grueling academy where they were evaluated and trained prior to being further mentored on the job by more experienced personnel. To a large extent, many municipal police departments subjected their recruits to similar preparation. County Sheriff, however, was generally an elected position for which prerequisite training in the methods and practice of law enforcement varied widely. Some of the best cops Horace had ever known were sheriffs, but, unfortunately, the opposite was also true.

Unbeknownst to Bobby Wade, the police lieutenant with whom he was now speaking had done his homework. Horace had solicited information from his colleagues about Wade and recent events in Taos County. Karen at dispatch had been of enormous help. On his drive north, Horace had stopped briefly at the station in Santa Fe. In the hour and a half it had taken him to get there, Karen had prepared and relayed quite a dossier on the sheriff. Competent, hardworking, and honest were positive traits attributed to the man, but he was also known to be obdurate, overbearing, strong-minded, and more than a little insecure. Begay was doing his best to appear non-threatening and helpful. At first Wade had been suspicious and reserved, but he gradually wore down as Horace worked his placating magic.

Bonifacio Romero, his wife Rosa and a young man named Greg Martinez milled around deferentially as the two lawmen conferred. Horace had cautioned them—particularly Boniface, a man with a short temper—about the politics behind their visit. All of them knew that good relations with the local constabulary were essential if they were to be of help to Denny.

Sheriff Wade finally approached the group with a smiling Horace Begay one step behind. "I appreciate the concern y'all have for Mr. Carswell, but you have to understand the gravity of what happened out there at that ranch. A man is dead, shot to death with Carswell's gun. He will remain our prime suspect until the legal system reviews the evidence. Now, Officer Begay assures me you might be of assistance in my investigation, so I'm gonna let you speak with the prisoner individually. Mrs. Romero, you can go in with your husband, but Mr. Martinez, you'll have to wait 'til they're done. I'll give each of you ten minutes." Satisfied that he had the situation well under control, Sheriff Wade nodded to Camille, bestowing permission for her to retrieve Denny from his cell.

Although he was embarrassed by his appearance, Denny was delighted to have visitors. He was heavily shackled, both hands and feet, because instead of meeting in his cell they were in a sort of interrogation room set back from the booking area. A clip fastened a chain that ran from the manacle on his right hand to one of the legs of a stout table situated in the center of the room. It provided a further means of control over violent criminals, but in his case Denny figured it was a subtle reminder from Sheriff Wade of who was in charge. Aside from the table, four chairs, and an overhead light covered by heavy steel mesh, there were no other furnishings in the room.

Rosa Romero sat with her back straight looking at Denny with a mixture of compassion and sadness. She was a petite woman in her late '70s. Despite her old fashioned granny glasses and grey hair, the twinkle in her dark green eyes made her look ten years younger. The one trapping that broke from her otherwise Victorian appearance was a fine silver chain necklace. It supported a beautiful hand-crafted turquoise crucifix with silver border, jewelry Denny had never seen her without.

Boniface was proof positive that opposites attract. In contrast to Rosa's reserved countenance and demeanor, he was petulant and temperamental, a short, ruddy-faced bull of a man who was offended by the harsh treatment being meted out to his friend. He had been Denny's mentor years ago, and the two had worked so closely together they almost knew one another's thoughts. At the moment Boniface's face was contorted in a scowl; his fists were balled tight and pressing on the table top.

"That son-of-a-bitch, shit-for-brains sheriff, how could he possibly think you had anything to do with killin' a man? You ask me, all he cares a rat's ass about is finding a scapegoat so he can carve another notch in his gun handle, something to wave around at election time."

Denny smiled; the tirade was expected. What surprised him was that it hadn't been delivered earlier, to Sheriff Wade's face. Boniface had never before, at least not to Denny's knowledge, held back when someone needed a good tongue lashing. He suspected Horace and Rosa had a lot to do with Boniface's good behavior. "So, Boniface, tell me how you really feel about Sheriff Wade."

The older man let out a deep breath, "Sorry, poor Rosa's been listening to me bitch all morning; don't know how she puts up with me."

"I understand," Denny said with a wink. His voice became serious again, "Thanks for coming up here to help me out. I can't convey how much I appreciate your support."

"We're your family," Rosa replied. "If either of us were in trouble you'd do the same."

"Of course," Denny said. "Look, we only have a few minutes. I don't have a firm grasp of what's going on yet, but there are a couple of things that need looked into. With your help we can gather enough evidence to either exculpate me entirely or to point the sheriff's investigation in the right direction."

"That might be tough," Boniface scowled, "I think he's pretty dumb."

"I have jobs for both of you," Denny continued. "Boniface, as soon as you leave here, while the sheriff is still in a good mood, I need you to analyze the copies of my field notes that he gave to me yesterday." Denny carefully outlined exactly what he wanted and why. Boniface nodded conspiratorially as Denny turned to Rosa. "Rosa, your job may be the most difficult and the most important. Augusto Benavides, the man who was murdered, had a wife who currently lives in a nursing facility in the nearby town of Arenas Rojas. I don't know if Sheriff Wade has questioned her at all, but she and Augusto were very close. I'm betting she might have information that can shed light on the killing. She reminds me a lot of you. She grew up in an isolated, rural part of the State and she had a sister with whom she was quite intimate. My hope is that the two of you may forge enough of a bond that she'll open up a bit, reveal something of importance. Her English is apparently not very good; you'll have to speak with her in Spanish." He looked over at Boniface with concern, "It may be dangerous for you, Rosa. The killer is still at large. Please take Boniface or Greg with you—and be very careful."

Camille knocked softly before opening the door. Rosa and Boniface said their goodbyes and were ushered out before Greg Martinez was allowed in.

"Hey, boss. Wow, you look like shit!" He reached forward and shook Denny's shackled hand.

"Thanks," came the sardonic reply, "two of your greatest assets, niño, are your honesty and tact."

"Yeah, I know. I've got some good advice for you too: fire your tailor. Even by today's standards those threads don't cut it."

Denny couldn't help but laugh. He'd forgotten how ridiculous he looked, sort of like a Buddhist monk in prison orange. "Point taken. Hey, thanks for helping me, Greg. I'd like to chat more and find out about the work you're doing for me in Santa Fe, but it will have to wait until we're done here.

"First of all, I'd like for you to double check the calculations Boniface is doing. It's not that I don't trust his work; hell, he's the guy who taught me everything I know about field astronomy, but a lot depends upon those results. If my hunch is correct, they may put us on the killer's tail. I'd also like you to take a look at the survey data that pertain to the Stetson ranch. My copies are being held as evidence, so you'll have to go across the street to the County Clerk's office and look at the official documents on file. The verbiage in the older deed, the one authored by Antonio Roybal back in 1992, seemed to me to be inconsistent and/or incomplete. His is the description that actually defines Stetson's parcel—what was sold to him by the Benavides family. A second survey, done by a surveyor named Josh Marquez, was done later in an attempt to clarify the earlier description. I'd like to get your thoughts and those of Boniface about those two surveys. Maybe you can discuss the documents with him before he and Rosa take off for their interview with the murder victim's wife."

Greg listened attentively while Denny talked. His lighthearted manner was dissolving away as Denny's urgency and the reality of what they were doing sank in. He noticed how much older his boss seemed since they'd last spoken less than a week ago. There were circles under his eyes, fatigue in his voice and he smelled like he hadn't showered in days. Denny was going through hell. The more Greg listened, the more admiration for his mentor and friend welled up within him. The man was a fighter. Greg had little doubt that their efforts would exonerate Denny and perhaps nail the real killer. He was proud to do his part. But how else could he feel? The consequences of failure were too awful to contemplate.

Horace Begay slipped into the interrogation room after Greg left. "Fill me in, Denny. Tell me what happened and what you've asked Boniface, Rosa and Greg to do—nothing illegal, I hope."

Denny smiled with enormous relief. He no longer felt alone and abandoned. People were helping him now, competent people in whom he had faith and trust. Horace had driven all the way up from Los Lunas to render assistance, and his positive influence on Sheriff Wade was a godsend. Denny felt his situation was much improved over what it had been just one hour earlier.

Starting from the beginning, he told Horace everything. The night before, he'd rehearsed what he wanted to say, trying to make it quick, thorough, coherent and concise. For twenty minutes he talked, giving details about the land sale, the deed and survey descriptions, pertinent findings from his own work and his interactions with Larry, Bella, Augusto, Emilio, Tommy Hawkfeather, Lily, Xeno and Josh Marquez. He described exactly what he'd seen through the scope of his total station and spent considerable time talking about the interviews conducted by Sheriff Wade at the murder scene. Lastly, he attempted to make sense of what was going on and what evidence he hoped Greg Martinez and the Romeros might find.

Horace listened attentively, nodding occasionally when Denny hit upon key points. When the narrative was finished, the room fell silent save for the background murmur of the desk sergeant talking on the telephone in the adjacent room. Horace emerged from his reverie.

"You may not like it, Denny, but we have to keep Sheriff Wade appraised of everything we find."

Denny objected strongly. "I don't think that's a good idea at all, Horace. Wade's shown no interest in my predicament. As far as he's concerned, the crime's been solved. In fact, I think he'll try to torpedo any further investigation into the truth of what took place." He added dolefully, "He's convinced he's caught the killer."

"The Sheriff has other things on his plate right now, Denny. He's not ignoring you so much as he's overwhelmed by another case."

Anger resonated in Denny's voice. "I'm on trial for *murder* here. There can be no excuses for his lack of attentiveness."

"Oh, I'm not offering any excuses for the way the Sheriff is behaving, only an explanation. You obviously aren't aware that another killing occurred in Taos County a few weeks before Augusto's death. A young boy about eleven years old was brutally knifed and his body dumped into the Rio Grande gorge."

The questioning look on Denny's face indicated that, indeed, he was unaware of the earlier crime.

"It gets worse," Horace continued. "The boy was a good kid, a member of the Taos Pueblo. The tribe is understandably upset about what happened, and they've lit a fire under Sheriff Wade. Their officials have been rather vocal, especially of late, offering a large reward and venting their frustrations to the press. News articles are accusing Wade of racism because of his perceived lethargic investigative response. Ninety percent of what the two of us talked about when I arrived this morning dealt with that issue. He's essentially moved your case to the back burner."

Denny's ire abated somewhat. It was difficult for him to feel much sympathy for the Sheriff, but Horace's words did carry with them a new perspective and level of understanding. "So, do you think Wade will appreciate efforts on our part to investigate Augusto's murder?"

"If things are presented right, yes. As long as Sheriff Wade believes he's in control and has the opportunity to sanction whatever we do, I think he'll be very cooperative."

Denny nodded. Begay was a smart and reasonable man. His assessment of the Sheriff was born of a thorough and objective analysis. There was something else Horace said that stuck in Denny's mind. It reminded him of a statement Roy made during his loquacious rant the night he'd been brought in. "Did you say a reward was offered by the pueblo for information related to the boy's murder?"

"You know something?"

"Maybe." He looked Horace in the eye and told him that Roy's incarceration resulted from a fight he'd had with a colleague.

The altercation had presumably started with a disagreement about splitting a reward. "It may not mean anything, Horace, but one never knows." He smiled, "we convicts hear things in the slammer." He became serious. "When you talk to Sheriff Wade, please make sure he understands that I didn't know anything about that other murder until today. He has to know that I would never intentionally withhold evidence about something like that, no matter how pissed at him I might get."

"Of course." Begay grabbed his hat and stood to leave.

"Oh, Horace, there's one more thing you can do. One thing that's nagged at me ever since Augusto's murder is: what brought him to the summit of Cerro Medio that day? From what Emilio told me, Augusto spent most of his time with his wife at the nursing home. Someone must have contacted him, the killer most likely, to arrange a meeting. It was a perfect set up. No one should have been anywhere near the site at that late hour, and if you knew where to find a .22 rifle owned by an old surveyor, he'd be easy to frame for the murder."

"But the killer didn't count on that very same surveyor witnessing the shooting through a high-powered scope," Horace mused. "So, you think a check of Augusto's phone records might be in order?" Denny nodded in reply. "Well," said Horace as he opened the door, "I'd best get to work. I'll be by again this evening. I trust you'll be here when I get back." He grinned and left his friend behind, chained to a table but otherwise in excellent spirits.

BY THE TIME GREG got to the coffee shop, Boniface was already hard at work. Rosa sat quietly sipping tea, watching her husband flip through the pages of a dog-eared book and jotting down numbers. Denny's field notes lay spread out on the table along with a rough sketch, made by Boniface, of the traverse Denny had run two days before. So absorbed was he in his task, that Boniface failed to notice the new arrival for several minutes. When he finally did, Greg only rated a quick glance.

"Pull up a chair, Greg. I want you to check my math."

Greg sat and watched until Boniface came to a stopping point, whereupon the old surveyor gave him full attention.

"Use this almanac to find the Sun's Greenwich hour angle and declination at these two times on October 28th; that's two days ago. When you're done, use those data to compute the corresponding altitudes and azimuths of the Sun. The latitude and longitude of the Stetson ranch are right there in Denny's notes." He tapped on the photocopies and then immersed himself back in his work.

Greg was hungry. Before going to his truck to retrieve his calculator, he signaled a waitress and ordered breakfast: three eggs scrambled, three pancakes, bacon, home fries, coffee, and a large orange juice.

"My goodness," Rosa remarked once the waitress was gone, "you must not have eaten in quite a while."

"I'm starving," he acknowledged. "Half a dozen donuts are all I had for the drive up here this morning. That was almost **two hours** ago! I've eaten absolutely nothing since then."

When Greg was gone, Rosa affectionately took her husband's hand, "He reminds me a lot of you at that age," she said with a nostalgic sigh. "I remember your idea of a hot date was: 'Let's get a bite to eat.'"

Boniface grunted, "Yeah, I remember those days," he smiled and squeezed her hand, "but don't worry, Denny warned me about Greg, told me not to offer to buy him lunch or my wallet would float away."

Greg soon returned, calculator in hand, and got down to business. "You know, Mr. Romero, I've programmed my calculator to solve the astronomical triangle. Given any three parts, the code will solve for the other three. It will even spit out all combinations for ambiguous cases, like when you know two adjacent sides and a non-included angle—really speeds things up."

That's great, Greg. I've got similar routines that I use on my calculator and on my PC. But," he paused, "there are two important things you should know about me and your boss, Denny. Firstly, we represent a different generation, one that for years did exacting work without the likes of laser-based EDMs, optical encoders, electronic calculators and computers, and, most notably, high-precision GPS. Sometimes it's good to spend the extra time and use the old tools, like this set of log tables, just to reassure yourself that you still have what it takes."

Greg nodded. Denny had said much the same thing to him on numerous occasions. They'd even run a traverse once using an eighty-year-old transit and a steel tape so that Greg could appreciate the skills now being obviated by sophisticated new techniques. "So what's the other important thing I should know? You said there were two."

"Oh, yeah, I wanted to tell you that memory lapses occur frequently in one's old age. I have to use these log tables 'cause I left my damned calculator at home!"

Rosa watched with interest as the two men worked over Denny's data. Once Greg had eaten and the calculations were complete, the three of them sat drinking hot beverages. Boniface peered over his glasses in intense concentration as he studied Greg's results. "Very nice," he murmured, "Denny taught you well. He's going to be very interested in these results too, very interested indeed."

"I don't really get it," Greg said. "I mean, I understand what we did to find the Sun's altitude and direction, but Denny didn't make any solar observations. What good does it do to know the Sun's position if you can't use it to determine the bearing of a line?"

Boniface smiled wryly, "The traverse data tie in to several GPS points; we already have the bearings we need."

Greg was now even more perplexed, "So, why the heck are the solar results important?"

The glint in Boniface's eyes intensified, "You know what a sundial is, right? Well, let me tell you what's going on . . . ."

THE WOMAN MANNING the desk at the nursing home suggested they stop by at 2:00 PM. Lunch would be over and all of the "inmates" settled in their rooms by then. Over the phone, Rosa learned that Maria Benavides was one of the more popular inhabitants at the facility, had made lots of friends. She had recovered amazingly well from her stroke and was now more of an asset to the staff rather than a patient. She could walk with the assistance of a cane and was in the process of preparing to move back to her home when Augusto was killed. The receptionist went on to say how supportive everyone at the facility had been in her time of need.

With the trip to the nursing home delayed, Rosa decided to spend a few hours shopping among the boutiques in Taos. Tourist season was over and the ski crowd had yet to arrive. That made it the perfect time to hit the specials and sales that accompanied the merchants' temporary slump in business. Boniface and Greg opted to spend the time doing research.

THE ELDERLY GENTLEMAN at the County Clerk's office looked up from the book he was reading when the two surveyors walked in. His dour expression resulting from this interruption turned into glee when they told him what they wanted.

"Let me search our computer system," he announced brightly. Swirling his mouse across a worn pad by his right elbow, a large Sun Systems CRT monitor tethered to an ancient IBM PC slowly came to life. At the command prompt he carefully typed the name Stetson and hit 'return.'

I don't believe it, Greg thought in amazement, *he's running a DOS operating system. These guys aren't even in the Windows era yet!* He glanced over at Boniface who was peering over the man's shoulder, looking at the screen. *That's almost like using log tables to do survey calculations.* He was suddenly grateful that Boniface couldn't read his thoughts.

"What you need is in deed book numbers 574 and 580," the man announced proudly. "Both should be in aisle two, towards the back."

There were two rows of property records lining either side of a short metal bookcase. Conveyances and supporting documents such as survey records, easement grants, etc. were neatly arranged chronologically in heavy three-ringed binders called 'deed books.' Great care had been taken to punch holes in the pages of every document so as not to deface any information it contained. Fabric reinforcing disks had been glued around the holes to prevent tearing, some of which had been trimmed in order to keep things legible. Cross references to older, parent deeds were recorded neatly in pen on the first page of each document. One could tell the old filing/referencing system was still maintained even though a "modern" computer database was also in place. Greg was impressed; the employees who worked in that office took real pride in preserving the land records of Taos County.

Boniface took book 574, the one that harbored the Roybal description, while Greg took the one with Josh Marquez's survey. Taking seats opposite one another at a small table, both perused their documents under the watchful eye of the old man at the front desk.

For the life of him, Greg couldn't see anything wrong or out of place in the Marquez report. The verbiage was clear and to the point, and the accompanying plat agreed with the text in every detail. Convinced that the report was in order, he sat back and waited for Boniface to finish with the Roybal document.

Boniface spent considerable time reviewing the original Stetson deed and survey. Ten minutes went by before he spoke up. "It bothers me that this fellow Roybal didn't include a plat in his description. Not only that, the last metes and bounds courses sound disjointed. I agree with Denny; something's not right. Tell me what you think, Greg." He passed the book over to Greg and then stood to stretch his legs.

Greg breezed through the "legalese" portions of Antonio Roybal's report before concentrating on the survey section. Everything looked pretty standard to him until he turned to the last page. Even with his comparative lack of experience, he had to admit to Denny and Boniface's conclusions. Flipping back and forth he read and re-read the last two pages. He was about to close the book when something strange caught his eye. He stared for a moment to make sure he wasn't mistaken before blurting out:

"Holy shit, there's a page missing! Boniface, come here; look at this."

The outburst drew a disparaging frown from the fellow at the desk. Boniface hurried to where Greg sat and looked over his shoulder.

"Up here in the corner, see that?"

Between the pages, and underneath the staple that joined them, was a small sliver of paper. Fibers protruding from the shard indicated it was a remnant of a page which had been torn from the document.

"I'll be damned," Boniface replied, "you've got sharp eyes, Greg. Most people never would have noticed that."

XI

Boniface and Rosa arrived at the nursing home in the small town of Arenas Rojas just before two o'clock. Two ancient looking women and a man who looked old enough to be the father of either one sat in chairs on the porch of the 'Red Sands Retirement Home.' Why the English equivalent of the town's name graced the sign out front had been a subject of much speculation since construction of the facility fifteen years earlier. One of the crones looked up as they mounted the porch steps and smiled in toothless greeting. The front door opened into a communal area that surrounded a circular reception desk/nursing station. Straight beyond the desk was a small dining room, and past that was the kitchen. A single transverse hallway led to rooms where the patients lived.

"Good afternoon, you must be the Romeros," an affable woman in blue hospital garb said as they approached. "Maria is looking forward to having company. I'll have to warn you though; she's had a rough time since Augusto's death." She led them down the right passageway and stopped outside room 23.

"Come in," Maria answered when Rosa knocked. It was as though they crossed the threshold into a different world when they entered Maria's room. Instead of 'institutional white,' the walls were painted in a cheerful almond-bisque. Long, colorful drapes hung around the one window, pulled back now to allow in sunlight that filtered through the branches of two majestic cottonwoods out in the building's courtyard. Pictures of relatives and of what appeared to be the Benavides ranch adorned the walls and bureau. Upon a stout table opposite the bed sat a large wooden shrine consisting of a decorative base atop which stood a statue of 'Our Lady of Guadalupe.' The base was covered with wax drippings, and a votive candle flickered softly in the air currents generated by the room's inhabitants.

Introductions were conducted in English for the benefit of the nurse, but when she left, Rosa switched to Spanish. During the drive from Taos, she had rehearsed her introductory remarks. Through the use of old New Mexican colloquialisms and nuances in pronunciation, she hoped Maria would recognize their commonalities in background and culture. Maria's face transformed as Rosa spoke; the plan worked well.

"Oh, you're from Pecos," Maria cooed, "I used to ride there through the mountains with my sister and her friends when we were in high school. You must know the familia Cordero then. Luis was such a handsome man, no? My cousin, she dated him once."

"Yes, Rosa replied, "but I'm a few years younger than you; Luis was Alcalde when I was in school. But he was certainly handsome, you're right about that."

The conversation remained convivial for several minutes before Rosa turned to business. "I wish this were more of a social call, Maria, but I'm afraid the tragedy with Augusto is what brings us here. It's terribly sad; if anything like that ever happened to Boniface, I just can't imagine . . . ." She choked back a sob and wiped tears from her eyes. "We want to catch whoever did that, Maria, and we think you may be able to help us. A good friend of ours was there that night. He was surveying the portion of the Stetson ranch that the pueblo wants to buy and he saw someone . . . arguing with Augusto."

Boniface broke in, "Denny, the surveyor, didn't harm your husband, ma'am, but he's in prison for the crime." He leaned forward, his voice tense, "Even though Denny's in jail, he's figured out that there's something wrong with that land deal. He suspects that Augusto's protestations were valid, and that's what led to his murder. There are a couple questions we'd like to ask if that's okay." He stopped to monitor her countenance. Maria's eyes glistened, but she met Boniface's gaze and nodded.

"That afternoon, did Augusto tell you why he wouldn't be here with you, because he had to meet with someone?"

She shook her head, "No, we spent time together in the morning. It was such a beautiful day, 'muy linda.' We walked around the building and visited friends. He said he would be back, but he never came."

"Had he said anything to you in the past week or so about trying to stop the land sale? When Denny spoke to him, Augusto mentioned he was going to contact a lawyer and a surveyor in an attempt to obstruct the sale."

"Oh, he was always saying that, but he would not have done so; we don't have the money. I don't remember much about the land we sold to Dirk many years ago. I've been sick you know; there are so many things I just can't recall. It is very frustrating. Augusto kept saying we never sold Dirk that land. That's why he was so upset. He hired a surveyor several years ago to try and stop Dirk from putting a fence on our land, but the judge, he ruled against us. I kept telling him, 'it doesn't matter now, Augusto. We are old. What do we need the land or the money for?'" She looked over at the devotional statue, made the sign of the cross and sighed, "It is in God's hands now, no?"

Rosa took Maria's hand in hers. "That's right, Maria. Whatever happened was God's will. Everything has a divine purpose." She gave Boniface a stare that indicated, "She's had enough—we should leave her alone."

Her husband nodded and they stood to leave, but Maria spoke again, "It's not the first time people tried to take our ranch. I'm ninety-one years old now and it's such a strange thing: I have a terrible time remembering recent events, but I have no such trouble recalling my childhood. I can't tell you what I had for lunch, or even if I had lunch, but some of my most vivid memories are stories my father told me about the land we live on. His father and grandfather fought the 'ladrones,' (the thieves) who tried to take their land. Papa was only six or seven when my abuelo died, but he told me how happy abuelo was when the ladrones lost their jobs and couldn't bother us anymore." She had a triumphant look about her as she reminisced about her lost relatives and past events. "So you see, it has not been easy, but God has helped my family stay here for many years. I think maybe this latest problem is simply punishment for having sold that land to Dirk. It's in the Bible, no? 'The Lord giveth and the Lord taketh away.'" She again crossed herself and settled back in her chair.

Boniface was lost in thought. Rosa nudged him in the ribs as she handed him his coat. "We'd best be going now, Boniface. Maria has been so kind to talk to us. We shouldn't wear out our welcome."

Boniface wasn't listening. "Maria, who were these 'ladrones'? In the clerk's office I saw that it was your grandfather who bought your ranch land from the Government back in 1905. He owned the land outright. Who tried to take it from him?"

The question puzzled Maria. She shrugged, "The ladrones were from the Government. My father said they didn't like the papers."

"What papers?"

"The old papers, the ones that tell who owns the land."

"Hola, Tia." A well-dressed man suddenly appeared in the doorway. His face was flushed and he glanced furtively back and forth between Maria and the others. "I'm sorry if I'm interrupting."

"Oh, sobrino, it is always good to see you. You are never an interruption," she replied in broken English.

"Who are these people, Tia?"

"These are old friends, Conrad," she said quickly. Turning to Rosa and Boniface: "This is my, how do you say, niece—no nephew—Conrad Sussman. Conrad, these are the Romeros. They are old friends," she repeated, "but they are just leaving." She became even more pensive as she reverted back to Spanish and addressed Boniface's question. "Augusto, he said it was bad to show the old papers to anyone, but he also said it was okay to show people a copy. I think there is one in the house."

Conrad and Boniface exchanged suspicious looks as the Romeros left. On the walk to the car Rosa asked, "You don't seem to like that man. Who is he?"

"That, my dear, is the asshole lawyer who Denny fired yesterday. Horace told me about him on the phone last night."

"Oh my, things here really are complicated, aren't they?"

"We have to talk to her again, Rosa. I want to see those papers she told us about. My hunch is that she's very confused. This whole mess may involve more than the sale of a few acres of land on which to build a damned casino."

THE MORNING HAD gone well for Denny Carswell. He'd gotten a good night's sleep, partly because the cavalry had arrived in the form of Horace, Boniface, Rosa and Greg, and also because circumstances had finally taken their toll. His stress level over the past few days had only allowed him a total of a few hours of real rest. Camille, bless her heart, had arranged for him to take a much needed shower, and with great ceremony she had then presented him with a clean jumpsuit that was only one size too big. Under the circumstances, life was good. After slogging through a bowl of sticky oatmeal, he felt almost human.

At half past eight there was a commotion up at the front desk. An hour later, after events had settled down, Horace Begay stood outside Denny's cell. He was alone. Denny looked at him, his brow elevated, anticipating news.

"According to Boniface, your hunch was right. You'll have to go over the specifics of the calculations with him and Greg, though. Also, Greg found this." He held up a plastic bag containing a document. It was the copy of Dirk's deed, the one from the Clerk's office. The pages were folded back to reveal the last Page of the Roybal survey description. Pushing the bag up close to the cell window, he pointed to the staple in the corner. "Look closely and you can see the remainder of a page that was torn out. This is the official copy that was on file in the records office. When they told the Sheriff about it, and its potential significance, he immediately bagged it and put it into evidence." Horace stood silently while Denny contemplated the new information. "There's more," he continued. "There were two calls involving Augusto's phone the day he was killed. One was an outgoing call made in the morning to his wife's nursing home. The other was an incoming call received shortly after 1:00 PM. The call came from the 505 area code, but the number was untraceable; it probably came from one of those cheap pay-as-you-go phones. It could be that the call was legit, from a friend, but my guess is that it was from his killer."

"Someone he knew," Denny muttered, "or he wouldn't have been willing to meet the caller at a remote site later in the day. Is there any chance you can track down that phone?" He wasn't very optimistic, but he had to ask nonetheless.

"No, but if we find it, the number will be programmed into it. Whoever possesses it will have some serious questions to answer."

"I'm a bit surprised Sheriff Wade thinks the issue with the deed is enough to classify it as evidence," Denny said, changing the subject.

"The Sheriff's mind is beginning to open up," Horace said. "I think the arrival of your entourage, meaning me, Greg and the Romeros, and the support we're providing to you, has him willing to entertain the possibility that you might not be Augusto's killer. Also, I can tell he's curious about motive—wants to know why the murder occurred, not just who committed it." He tucked the bag containing the deed under his arm and regarded Denny and his surroundings with a hint of disgust. "If it were up to me, I'd open this door, let you out and apologize for all you've been through. But that's not going to happen. You're still in trouble, Denny. The Sheriff keeps harping about evidence—evidence this and evidence that—and the fact that it was your gun that was used to kill Augusto, in his mind, overrules all else. Despite his willingness to accept additional evidence, he still thinks you're guilty. The ballistics guys at the State forensics lab did me a big favor and ran expedited tests on the slugs from Augusto's autopsy. It was your gun all right. I'm afraid that's reinforced Wade's suspicion about you. If our investigation stalls, you're going to have to find yourself a good lawyer. I'm sorry, but that's the way it is."

Denny was barely listening. Horace Begay was a man of few words, and the length of the man's homily, much more than what he said, testified to the gravity of his predicament. Denny's thoughts were wandering. He was trying to process the information garnered by Boniface and Greg.

Horace let out a deep breath. "I gotta go. They brought in Jimmy Chung earlier this morning. The Sheriff's keeping him in the women's cell block so that he and Roy can't talk to one another. That case is about to break wide open and I've been roped into it." He smiled for the first time since his arrival, "What you heard from Roy that first night is paying off in spades for Sheriff Wade. That may buy you some consideration in the future. We can only hope."

Denny asked, "Do you think this fellow, Chung, knows who killed that boy?"

"Yeah, he knows who did it, but he's playing coy. I've got to hand it to the Taos cops. They know the guy, and when Wade got a warrant for his arrest, instead of going to his home they staked out one of his known places of business. They caught him red-handed, swapping 'unauthorized pharmaceuticals' for stolen goods. With felony distribution charges and parole violation hanging over him, he has no choice but to cooperate fully if he doesn't want to spend most of his life in jail. Right now I think Sheriff Wade would kiss all four of Jimmy's cheeks, drop all charges, and buy him a steak dinner if he gives up the killer."

Denny quickly shifted gears as Horace began to leave. "Wait, Horace." The officer waited patiently while Denny formulated what he wanted to say. He seemed to be staring at nothing, lost in thought. Finally, he continued, "You have to come back, Horace, and when you do, bring Greg, Boniface and Sheriff Wade with you. No, wait, on second thought, don't bring Boniface, he'll just get into it with Wade, but be sure to bring the other two."

Begay smiled, "Any further orders, Mr. Prisoner?"

Denny shook his head, "I'm sorry, Horace, but this is extremely important, and," he added with bitterness, "I'm sick of sitting in this cell."

The policeman nodded in understanding. "We'll all be here. You can bet on it."

When Begay was gone, Denny lay on his cot and stared at the ceiling. The torn page from the deed was a godsend. *It doesn't really matter what is on that page,* he thought. *It's given me what I need to finally get out of this shit hole.* "You've been clever," he murmured as though speaking directly to Augusto's killer, "taking my gun, setting me up the way you did, to both eliminate Augusto, who'd become a problem, and to stop me from perhaps discovering the truth about who really owns that land. But you've made at least two big mistakes now, buddy, and I've got the upper hand." He clenched his fists. With uncharacteristic malice, he added, "It's time to turn the tables, time to make *you* sweat for a change, you Bastard!"

AS PROMISED, Horace assembled the participants requested by Denny for a brief meeting. Greg was anxious to find out what Denny had up his sleeve and had arrived at the jail a half an hour early. He was now impatiently waiting for the conference to start. The Sheriff had been a problem for Horace. It had taken all of his persuasive power to finally get the man to agree to Denny's request. As with many things, the Sheriff's ego considered this to be a threat to his authority. He was busy trying to apprehend the Indian boy's killer, he'd told the State policeman; he couldn't spare any of his precious time to satisfy the whims of a 'gol-darned' prisoner. Horace's anger had flared at the derogatory reference to his friend: the thoughtful, innocent man whose entire future was in jeopardy. It had been a struggle to maintain his composure and press on. Denny was right about Boniface, Horace had mused, if the Sheriff had made a comment like that in front of the old surveyor, fists would have flown. That would have made matters much worse—having to deal with two friends in jail and an angry and even more cantankerous sheriff. Finally, Wade had given his consent when Begay reminded him of Denny's important tip; the one leading to the arrest of Jimmy Chung. When Horace put it that way, it made it seem like a deal—a prisoner providing valuable information in exchange for a moment of the Taos County Sherriff's precious time.

"Aw right, let's get on with this. I ain't got all day," Wade said as he plopped into a chair on the opposite side of the table from Denny. They were in the interrogation room, the same one where Denny had been allowed to meet with Boniface, Rosa and Greg the day before.

It was all Denny could do to keep his cool. He stifled an all-consuming urge to tell Sheriff Wade what a pompous ass he was. That sentiment would have to be shelved until this ordeal was over. Exercising world-class self-discipline, he thanked Wade profusely for agreeing to meet with him. "Believe me, Sheriff, I wouldn't have asked for your time if this weren't really important." Horace looked on, straight-faced and more than a little impressed at Denny's self-restraint. "We're all aware that the file copy of the deed to the Stetson ranch is missing a key page from the surveyor's report. I'm certain that whatever is on that page provides the motive for Augusto Benavides's murder. It is important for the killer to keep it secret. Although I don't know what the missing page says, I can make a pretty good guess. Also, from the calculations Boniface and Greg worked up from my survey data, I'm convinced it was someone present at the Stetson ranch the day Augusto was killed who pulled the trigger." Sheriff Wade was listening, but the look on his face conveyed skepticism mixed with boredom. Denny decided he'd better get right to the point. "I have a proposal for you, Sheriff. You know how a rodeo rider is required to spur his mount to make him fight?" The analogy was poor, but it was all Denny could come up with on the 'spur' of the moment. Emilio had mentioned the Sheriff's rodeo exploits, so he decided that not-so-subtle pandering to the man's enormous ego might prove effective. "Well, our killer is like a bronco eatin' hay in his stall. He's calm and secure, thinks he's gotten away with murder. Well, I think we can stick a spur in his rump, make him mad so he'll tip his hand."

For the next five minutes Denny outlined the rudiments of his plan. He then turned to Greg, "You're the key, Greg. You have to be subtle, yet convincing. The new business cards I printed last winter have your name on them. That'll help a lot. In addition, although the Stetson and Benavides properties both lie entirely within Taos County, they also border Mora County. That's fortuitous; it too will help with the deception. How long will it take you to prepare the paperwork?"

Heck, it's only an hour's drive to my house," Greg replied enthusiastically, "I can be back here in four hours."

"There isn't that big a rush on this. Take your time and do it right. Tomorrow morning is plenty soon enough."

Sheriff Wade interjected, "The Mora County Sheriff will have to be involved. I'll call her and set things up."

"That's an excellent idea," said Denny, "I hadn't thought of that."

Horace Begay covered his mouth to "cough." He was actually stifling a laugh. Denny was playing the Sheriff well, leading him to draw the conclusions he wanted. Like a fisherman deftly presenting a fly to a wily trout, he'd teased, and finally hooked, the Sheriff's interest. The man was even growing excited about the plan.

"I also want to talk to Judge Hinojosa," Wade continued, "If this leads to an arrest, it's got to be righteous."

Denny nodded. "That's another good idea." He and Horace exchanged surprised looks. This time the Sheriff's input was not only important, but unexpected.

The meeting disbanded and Denny soon found himself alone in his cell. He was enormously relieved at the Sheriff's willingness to go along with his plan. What would he have done otherwise? Moreover, how could he even have attempted to extricate himself from his predicament without the assistance of friends like Boniface, Rosa, Greg and, especially, Horace, a decorated and respected member of the State Police? *I should be counting my blessings,* he reminded himself. *I have friends, surveyors, who might have been hanged after innocently accepting that job at the Stetson ranch. Heck,* he realized with a twinge of paranoia, *that might still be my fate.* Deep down, however, Denny felt a swell of optimism. Sheriff Wade had come around in that meeting. The man's conscience, or at least his curiosity, was finally winning over his gigantic ego. The two of them had achieved a rapprochement and were now working together in a meaningful way. It was all Denny could really have asked for and more than he had expected. As a result, he knew the investigation had turned a corner. The situation was much improved. He reveled in the idea that Augusto's killer was about to have a very bad day.

XII

The young man stood in the doorway of the Stetson ranch house. Tall, handsome and dark-haired, he had a friendly, easy-going manner that always put those around him at ease. Although Hispanic, his eyes were a striking azure blue, an unusual trait passed on to him and his siblings by his father. "Hi, I'm Greg Martinez." For good measure he handed a business card to the man who opened the door:

> Carswell Surveying
> Dennis Carswell, PS, lic. No. 592864
> Greg Martinez, Survey Technician

"Hello, Greg. This is a pleasant surprise," Larry Cooper said as they shook hands. "We didn't expect to make much progress on our land sale this week given the horrible events that have transpired."

Greg nodded gravely, "Yeah, I finally got a chance to speak with Denny, my boss, yesterday. He's in a real bind. Of course, you know more about that than I do. "

"Indeed, it's all very disturbing. If it helps any, Greg, I personally don't think he's guilty.

"Thanks. Anyway, like I told you on the phone yesterday, Denny asked me to come up and finish the survey. The pueblo is applying pressure and he needs the money to help with legal expenses. So," he smiled and waved in the direction of his truck, "here I am."

Larry ushered him inside. "Would you care for something to drink?"

"What? Uh, no thanks. Holy cow!" Greg said as he looked around. The various images displayed around the room, of Dirk from his movies, had captured his attention. The photo immediately to his left had made the greatest impression. It was a close-up of Dirk shaking hands with none other than the greatest western personality of all time. At the bottom it was signed: 'Happy trails, Dirk. Your friend, John Wayne.' "I sort of hate to ask this," Greg said sheepishly, "but do you think I could meet Dirk and Bella at some point? It would be so cool to get autographs from them, not for me, of course, for my sister."

Larry laughed. "That can definitely be arranged. They both love talking to their fans, and I know that they'll be very grateful for the work you're doing on this land transfer. Dirk isn't here right now, he went into town for something, but Bella is around. I believe she may be out riding."

"Oh, that reminds me," Greg said, "Denny told me to get in touch with a couple of your employees, Emilio and Lily. He says they work in the stables and may be able to show me around the ranch, help me finish this job."

Larry winked, "Let's take a walk and see who we can find."

The two exited the house and headed straight for the stables. Outside, in the paddock, Lily was exercising one of the horses. She looked up as the visitors approached. "Good day, Lily," Larry cheerfully announced.

"Hello, Larry," she responded with a hint of distrust.

"Lily, I'd like you to meet Greg Martinez. He's here to finish our much needed land survey. Now, Greg, Lily doesn't always get up out of the right side of the bed, if you get my drift, but I swear I saw her smile once last year. Miracles do happen," he chided playfully.

"It's nice to meet you, Greg, and I'm not the misanthrope Mr. Cooper would have you believe." She glanced in Larry's direction and smirked, "I'm just not given to euphemistic behavior in the presence of those prone to constant and egregious peculation."

The genial expression left Larry's face, but he successfully fought to retain his composure. "Well," he said, attempting to maintain a lighthearted facade, “I'll have to consult my dictionary this evening to see if I've been praised or insulted. I suspect I know which it is, though, given that it's that ‘time of the month’.” Satisfied with his retort, and unwilling to wage a continued war of words with a skillfully armed opponent, Larry fell silent.

“Look, I'm just here to finish a survey job,” Greg said with a disarming smile, “ranch politics are none of my concern.”

“That's very wise,” Lily replied coolly. “So, what is it you require from me, Greg?”

“I was told to talk to you and Emilio. I'm led to believe that one of you can escort me to the top of Cerro Medio.”

“Sure thing,” she replied. “Come on, Emilio and Bella are in the tack area. I'll introduce you.”

The men followed Lily as she led her horse down the central corridor of the barn. The floor was covered by a thick layer of straw. It had the earthy smells of horses and leather, and the faint ammonia odor from fresh manure and urine wafted out from the stalls that needed cleaning. They paused briefly while she settled the animal into its stall. As they neared the equipment room at the far end, the sounds of light conversation greeted them. Bella was lounging against a bench nursing a cup of coffee while Emilio worked with an awl, fixing a frayed leather strap.

Both looked up as the three visitors approached. Bella took one look at Greg and broke into a seductive smile. She was dressed in skin-tight riding pants, tall black boots and a loose cotton shirt. A denim jacket provided enough warmth to take away the autumn chill, but was strategically buttoned low enough to provide a good view of her ample bosom. The look was carefully crafted to be part ‘Lady of the Manor’ and part ‘Victoria's Secret Farm Girl.’ Greg was about to become ‘cougar prey.’ She put down her mug, collected her hair behind her neck, secured it with an elastic band, and walked over to meet him. “My goodness,” she exclaimed, eyeing him up and down, “here's someone I've never met before, because if I had, I'd definitely remember *him.*” She studied his reflective vest and the plumb bob hanging from his belt, “Hmm, there must be some law that requires all surveyors to be drop-dead gorgeous.”

Greg blushed crimson and held out his hand, "My name's Greg Martinez, Ms. Starr, er . . . Mrs. Stetson, ma'am."

"Call me Bella, and it's definitely *my* pleasure." She clasped his hand in both of hers.

Greg was flustered. He tried to act nonchalant, but couldn't suppress his shyness. "I'm just here to do survey work to try and finish, um, the survey job that my boss came here to survey for . . . ma'am," he stammered.

She continued to hold his hand for a moment, reveling somewhat in his discomfort. However, not wishing to torment him too much, she finally let go and walked back to where she'd left her beverage. Her tone sobered, "Dirk and I are anxious to get that survey out of the way. It's nice of you to do this, Greg, in light of what happened to Augusto and to poor Mr. Carswell. I want you to understand that I know Denny is innocent. It shouldn't be long before he's out of jail, but if his problems persist, I'll be the first person to testify as a character witness on his behalf."

"Yeah, how's Denny doin'?" Emilio interjected. "Is there anything we can do for him?"

"No," Greg replied, "but I know he'll really appreciate your concern. What he needs right now is for me to finish this job so that he can get paid and then turn his attention to his legal predicament. It shouldn't take me long to close the traverse he was working on. Then all that's left is for me to stake out a boundary that's mutually agreed upon by representatives from the ranch and Santa Teresa Pueblo. Actually," he paused briefly, remembering something else, "the layout work will have to wait until Monday. Denny said the property deed reads funny. He's concerned that Antonio Roybal, the surveyor, didn't plat the property. Because the ranch lies along the Mora county line, there should be a copy at the Mora County Municipal Complex. I'll finish the survey this afternoon and pop over there tomorrow to check it out. Maybe that copy of the deed has more detail." He looked at Emilio and Lily, "Well, if one of you has some free time this afternoon, I would really appreciate a guided tour to the summit of Cerro Medio. Once I set a prism there, the work I need to do shouldn't take more than an hour."

"I'll be happy to show you around, Greg." Emilio offered. "Just give me a few minutes to finish up here. You go get your gear and I'll meet you at your truck."

Lily walked to the tack room door. "Good luck with the survey, Greg. It was nice to meet you." Addressing Emilio she continued, "I'll finish working Bluebell and then feed and water the rest of the stock."

When she was gone, Bella adopted a knowing smile. "My, my, one would have thought she'd jump at the chance to spend some time with a dashing young man like Greg."

Larry laughed, "That, my dear, is strong testament to Xeno's animal magnetism."

"Yes it is," she agreed with a wistful sigh. "You're very perceptive, Lawrence."

"Oh," Larry chortled, "our dashing young surveyor has something else on his mind, although at this point I suspect he's too embarrassed to ask." He pretended to whisper to Bella, but spoke loudly enough for Emilio and Greg to hear, "He told me earlier that he would like your autograph—not for himself," he snickered, "for his sister."

Once again, Bella set down her coffee mug and walked suggestively over to Greg. Standing close to him—very close—she pulled a pen and a notebook from the chest pocket of his vest and asked in a sultry voice, "So, what's your sister's name?"

Greg blushed and swallowed hard. "It's, uh, my sister? Um, her name is Diana, Diana Martinez."

She licked her lips and thought for a moment, then wrote: 'To Diana, and to that hunk brother of yours—Best Wishes, Bella Starr'. She replaced the pen and notebook and then leaned in to give him a peck on the cheek. As she pulled away, she ran her hand lightly across his buttocks, and then sauntered towards the doorway. Pulling her hat from a nail on the wall, she smiled and shook her head, "An autograph. Why, Greg, that's the *least* I can do to thank you for all the work you and Denny have done for us."

Greg looked after her wistfully as she walked away. When he turned to his companions, Larry broke into a hearty laugh and he could see that Emilio was smiling from ear to ear. Greg couldn't help but chuckle, "Damn," he said, "all she did was give me her autograph and I feel like we just had sex!"

"Yes, she generally has that effect on red-blooded carriers of the Y chromosome," Larry acknowledged while trying to catch his breath.

Greg gave an exaggerated sigh, "I'd better pull my thoughts from the gutter and get to work. I'll see you when you're done, Emilio."

EMILIO SURPRISED GREG by bringing along two saddled horses for the trip to the survey corner up on Cerro Medio. As they rode along, silently at first, Emilio watched his companion. He gradually picked up the pace before commenting favorably on Greg's riding skills.

"I grew up with horses," Greg admitted. "My folks kept several on our farm but sold them about ten years ago along with part of our property. It didn't much matter though; I've got an uncle, my mom's brother, who owns a ranch not far from us, just outside of Española. I go there a lot to ride with my cousins."

Riding in the presence of someone who knew and loved horses served to loosen Emilio's tongue. The two men talked comfortably about ranching, farming and life in northern New Mexico while Greg worked. He set up his reflecting prism and pointed it in the general direction of the ranch house parking lot, visible across the long slope of sage and juniper earmarked for sale to Santa Teresa. As they prepared to ride back down to the house, Greg asked his host about what had taken place at that very spot the previous weekend.

"Ain't much to say really. I prob'ly shouldn't talk too much about it—might piss off Sheriff Wade, but over there's where Augusto got shot." He pointed to a group of small boulders thirty feet removed to the southwest. "All's I could see once I came up here was that someone put a bullet into his brain. I saw the hole in his forehead 'fore it got too dark to see. There weren't much blood, so I guess he must'a died instantly. Your partner's gun was over behind them rocks. My guess is the killer took it from Denny's truck while he was out surveyin' and then used it to frame him for the murder." He shrugged as though his conclusions were obvious.

"But the Sheriff doesn't seem to believe that," Greg countered.

"Well, son, you're right o'course, but I'll tell you somethin' about Bobby Wade. I've known the man for, shit, goin' on forty years now. He's a good man, hard-headed as a mule—there's no doubt about that—but he always does what he thinks is right. Don't get me wrong, I think he's mistaken about Denny. When it comes to judging people, I've got good whacha-ma-call-it . . . intuition. Bobby don't have that, but if the evidence points away from Denny, he'll come around. Anyway, it's in the hands of the courts now; I just wish I could help find out who really did this." He paused and Greg felt that Emilio's frustration was genuine. "What I'm ashamed of is that I've been so busy with this danged movie business, I haven't been to see Maria since Augusto died. I've been meaning to go with Bobby to talk to her. He don't speak Spanish and her English ain't real good. I know she might be able to shed some light on this mess. It could help Denny's case."

Greg thought it best not to mention that Boniface and Rosa had already met with the widow. He was sensitive to not divulging anything about the investigative work he and his friends were up to, but he did appreciate Emilio's line of thought. "If you think it will help, I can go with you," he suggested. "My Spanish is good, and I'm familiar with the deed histories of these parcels."

Emilio considered the idea and nodded. "That sounds good. What say you and me pay her a visit this afternoon?"

When Greg got back to his truck, he pulled out his total station and a tripod, leaning the latter against a front fender. Before setting up to make his last measurements, he pulled out his cell phone and made a short call.

Deputy Paul Viera flipped open his cell and listened for a moment. "Thanks," was all he said before hanging up. He made a second brief call and then pocketed the phone. Deputy Camille Baldonado occupied the passenger seat of the police car. They were parked in front of the small, satellite sheriff's office in Arenas Rojas. "We're on," Paul said as he turned the ignition key.

THE TOWN OF MORA IS the county seat of Mora County, New Mexico, one of the most sparsely populated counties in the contiguous United States. Located about forty-five miles southeast of Taos, the town lies in a high wind-swept valley overshadowed by 13,000-foot-tall peaks in the adjacent Santa Fe National Forest. NM 518 is the only thoroughfare linking the remote town to the rest of civilization, and it is often impassible after heavy snowfall in the winter. The county municipal complex is a group of double-wide trailers located on an unpaved side street off of NM 518. Among the small offices in the first trailer are those of the County Assessor, County clerk, and a small office of the State DMV. The farthest unit from the main road houses the Sheriff's Department and offices of employees of the Magistrate court.

Celia Mondragon, Sheriff of Mora County, sat in her office looking out the window at the front of the other building. The call she'd received the night before suggested that today might provide more on-the-job excitement since—when? In the last month she hadn't done much more than to write a couple of speeding tickets. All were given to outsiders who felt that, just because her town was small, they could zip through at highway speeds. She normally didn't fine anyone unless he was going at least fifteen miles-per-hour over the limit, or unless children were on their way to or from school. That never stopped many of them from raising hell though, about how unfair her "speed trap" was, but, hey, rules were rules. The highway was well posted, the signs legible and fair.

About the only excitement she'd had this entire year had been when Zeke Luna's prized breeding bull got loose and headed straight into town. Celia had chased it in her patrol car until she finally cornered it in the alley between the Food Mart and the gas station. Her frantic call to Zeke resulted in a rather insulting and lackadaisical response. He was busy mending the fence the bull had broken, he said, but he'd be by in an hour with his trailer. With a growing crowd of equally offended onlookers standing behind her,

Celia immediately informed old Zeke that she would shoot the animal if he didn't get his f***g fat ass over there within fifteen minutes. Zeke showed up with three minutes to spare. She'd charged the feisty old rancher with creating a public nuisance and hit him with a $550 bill for damage to the grill and left headlight of her cruiser, caused by the bull before he'd calmed down. She recalled with satisfaction how the incident had served to solidify her authority and enhance her standing as Sheriff.

In her seven years in office, Celia had developed good "cop sense." She prided herself on knowing when to go easy on folks and just give out warnings and when to get tough—dishing out tickets or making an arrest. Aside from the occasional domestic disturbance or Saturday night fight at the town's only bar, major occurrences like the bull incident were rare.

Celia's father, Pedro "Pepe" Mondragon, had been Mora County Sheriff for nearly forty years. She had grown up in his shadow, in awe of the respect he'd received from the community and fascinated by the myriad of difficulties he had to contend with over the years. At an early age she decided that law enforcement was in her blood, and when Pepe's ulcers forced him to retire, she'd decided to run as his replacement. The vote had been close, but she'd won. Most people in town knew her and trusted her, not because she was Pepe's daughter, but because she'd already served as his deputy for a year. Some members of this hardy ranching community had undoubtedly voted against her because she was a woman, but they had come around quickly. In the last election she had run unopposed.

Outside, the weather was cold and blustery. The summer-like temperatures of the previous week had yielded to a cold front moving in from the west. Patches of sun appeared and disappeared as clouds skirted by accompanied by dust devils whipped up by the winds blowing unrestricted across the valley.

Celia stopped gazing out the window and occupied herself with the small task of organizing her desk. She was a woman of action, more like her mother—with a fiery Latin temper and type "A" personality—than her calm, diplomatic father. She always suspected, though, that Pepe was more high-strung than his manner portrayed, given his career-ending problem with ulcers. The crutch that helped Celia to control her emotions was chewing gum. Whenever she was nervous, out came the 'Juicy Fruit.' It had been 'Juicy Fruit' now for several years. She much preferred licorice-flavored 'BlackJack,' but it turned the inside of her mouth black. That didn't bother her in the least until she noticed that it earned her strange looks from potential suitors. In a town as small as Mora, a single woman couldn't afford to cause herself "relationship impairment."

It calmed her nerves to exercise her jaws when she became fidgety, but chewing gum also had the undesirable side effect of making her the principal customer of the town's only dentist. Normally she hated paying homage to the local tooth-yanker as much as the next person, but getting her teeth drilled by Dr. Hathaway wasn't entirely unpleasant. He was soooo good looking, movie star handsome, that she actually looked forward to her semi-annual appointments. When he looked down at her as she lay in the dental chair, with his gorgeous dimpled smile backed up by his wiry build, broad shoulders and reassuring demeanor, she felt like she'd died and gone to heaven. There was one slight problem with Doc Hathaway, however. He was married. Actually, there were two problems. Not only was he married, he was *happily* married to a beautiful and very sweet woman. In fact, the couple was so nice she wouldn't allow herself to hope and pray that they would divorce.

Smack, smack, smack—the sound of the punishment she was inflicting on her mouthful of 'Juicy Fruit' broke her from her reverie. She had to focus on the situation. Bobby Wade had told her he was too busy to come over, but that the job was very important. He asked that she do exactly what the Indian cop, Begay, wanted. "Well, shit," she thought, "if Bobby is too busy to show up in person, it can't be that big a deal." However, she couldn't ignore politics. For Bobby to ask for her help was unusual and flattering, and she'd readily agreed to do what he wanted. Besides, the State cop, Begay, was good looking—not that she'd ever consider dating a fellow cop.

THE OLD WOMAN climbed the courthouse steps with some difficulty. She paused briefly on each step to catch her breath, and when she reached the front stoop she rested and looked around. The scene around her was one of a typical afternoon in a small mountain town. There was no traffic to speak of, and there were only four parked cars visible: one across the street at the post office and three others, including hers, at the municipal complex. An Indian sat on a bench across from the courthouse. He wore a wool cap and a torn canvass jacket—the kind one often saw in Army/Navy surplus stores. At his feet was an old duffle bag, an example of a bygone style with two large looped handles. Huddled against the wind, he was probably waiting for the next bus to Raton, Espanola, Taos—wherever; it was of no concern to her. She gripped the handle of her cane tightly and hobbled slowly to the front door of the building. She was on her way to the Mora County Clerk's office.

The phone in Celia's office finally rang. The caller said simply that she was going to the store, and was there anything Celia needed for dinner tonight? "Son of a bitch," Celia thought, "Bobby and Horace were right about the woman." The call was a pre-arranged signal from the clerk to inform her that her "perp" was leaving the building.

Celia lept out of her seat and hurried outside. The woman had just negotiated the steps in front of the other trailer and was making her way across the dusty parking lot.

"Stop! Mora County Sheriff's Department," Celia yelled. "I have a warrant. Please let me see what's in your purse."

"Oh my, are you talking to me?" came the feeble reply.

"Yes, ma'am. I want to see the document that's in your purse."

An incredulous look crossed the old woman's face, "Why, of course, officer."

With her purse slung over her left elbow, the woman reached into it with her right hand. The angle at which the hand entered the purse caused the Indian to jump to his feet. In a blur of movement he darted towards the old woman, covering ground at an incredible rate.

Time slowed to a crawl for Horace Begay. He was experiencing his worst nightmare, the same one that bothered many cops he knew: that of bearing witness to someone in desperate need and knowing that he might not arrive in time to help. But this was no dream. This was real life. There was still time; the hand wasn't out of the purse yet. Horace ran hard, driving legs that were stiff from sitting for so long in the cold, pushing himself onward.

He was a split second too late. A mere two strides separated him from the two women when the gun went off. The sound was muffled by the handbag, but the bullet found its mark. Ripping through the leather wall of the purse, it struck Celia Mondragon in the upper abdomen. The shock of the report and the impact of the slug lifted her completely off the ground and spun her around.

Begay knocked the old woman down and grabbed for the purse, wresting it from her grasp. He pinned her to the ground and cuffed her hands behind her back. After quickly checking her for concealed weapons, he crawled over to the Sheriff. "Call 911!" he shouted to an employee who had just emerged from the courthouse.

Celia was on her back taking shallow, stabbing breaths. Her face was pale, contorted in pain. She tried to talk, but only low groaning noises came forth. Horace tore open her shirt and pushed Celia's hands away from where the bullet had impacted the lower left quadrant of her bullet-proof vest—the vest he'd insisted she wear for the arrest. Ripping apart the Velcro ties that held it in place, he examined both the underside of the garment and the soft flesh underneath. He uttered a huge sigh of relief. It was the first breath he'd expelled since tackling the would-be killer. The vest had done its job. The skin over Celia's lower ribs was unbroken, but signs of extreme trauma were already beginning to manifest. Her skin was turning red and an oddly placed bump indicated that a rib or two must have broken.

"You'll be okay," he said with relief. Celia nodded and tried to smile, but couldn't manage more than a grimace. "Don't try to talk, just concentrate on your breathing. Try to relax. You've got some damaged ribs." All the while he was tending to Celia, Horace kept a watchful eye on his prisoner. She was making Herculean efforts to stand, but with her hands fastened behind her back she wasn't making much headway. Horace stood and lifted her to her feet. He pulled off her hat and grey wig. "Bella Starr, you're under arrest for the murder of Augusto Benavides and for the attempted murder of a peace officer. I must advise you to remain silent until you can speak with an attorney."

A small crowd had assembled, and several individuals were asking if they could help Sheriff Mondragon. Begay pinned his State Police badge to his old jacket in order to identify himself as an officer. "The Sheriff is not hurt badly," he reassured them. "She appears to have bruised or broken ribs but she'll be fine. I want everyone to stay back. This is a crime scene. The only people I'm going to allow within twenty feet of the Sheriff are paramedics." As if on cue, a siren could be heard coming from the direction of the town's small hospital. Celia helped to control the situation by trying to smile and waving to the crowd, giving them a thumbs-up.

Horace retrieved his duffle bag from beside the nearby bench and extracted several plastic bags. He donned rubber gloves and carefully began to rummage through Bella's purse. The first item to come out was a snub-nosed Smith and Wesson .38 caliber revolver. He turned to Celia, "So, what's it like to get hit by a .38?" She winced and stuck her tongue out at him. He smiled, bagged the weapon, and turned his attention back to the purse. Gingerly he pulled out a hastily folded and powder-burned piece of paper. One of the corners had been torn away, a sign that the page had been removed from a larger document. It was the fake plat that Greg had prepared and added to a copy of the Stetson ranch deed, one made to look like what Antonio Roybal must have included in the original. Horace had brought the fake deed with him that morning. He'd explained its significance in a meeting wherein he'd gone over details of the sting operation with Celia and with employees of the clerk's office.

Begay finished bagging evidence and hung around while the paramedics tended to the Sheriff. Celia squeezed his hand as they prepared to slide her gurney into the back of the ambulance.

"Thanks," she managed in a hoarse whisper. "You, the vest saved my . . . ." Her voice trailed off as her breathing became difficult.

Their eyes met. Each entertained a sense of attraction that brought on a blush. Horace gently placed his hand on her shoulder. "Good luck," was all he could think to say. Then he remembered: "Oh, wait. Sheriff Mondragon, I request permission to transport the prisoner and evidence related to her crime to Taos County."

"Permission approved," she murmured just before the ambulance doors closed behind her.

XIII

Greg Martinez sat in the parking lot of the Red Sands Retirement Home, sipping a Coke, waiting for Emilio Sisneros to show. He'd just rung off from talking to Boniface about the earlier visit the Romeros had paid to Maria Benavides. Greg knew the old surveyor was intrigued by what Maria had told him. She hadn't shed any direct light on who might have killed Augusto, but her claims to possess old documents proving ownership of her ranch property, papers that dated back at least one hundred years to the time of her grandfather, intrigued him. "There's more to this business than just that pueblo land deal," Boniface had said. "I'm looking into something else right now, but Rosa and I are going to talk to Maria again. If you and this other fellow, Sisneros, find out anything about these mysterious papers, let me know."

Greg slurped the last of his soda and set the cup into the plastic drink holder attached to the dashboard of his old Ford F100. There was a lot to think about. He figured Denny couldn't possibly be right about Bella. Part of the reason for his current visit to see Maria was to gain insight about the real killer. Closing his eyes he could smell Bella's perfume and feel, again, the sensuous way her hand brushed across his backside after signing his notebook. *Such a beautiful and seductive woman,* he thought, *a famous movie star who has it all. What could possibly induce her to become involved, in any way at all, in the pre-meditated murder of an elderly neighbor?* Well, he'd done exactly what Denny had asked. He'd surreptitiously misinformed everyone at the ranch about the fake deed he'd prepared. Whatever happened in Mora today as a result of that effort should clear Denny's name. "Personally, I think it's that guy, Larry, who killed Augusto," Greg muttered, "he seems like a real weasel."

A ruby-red sports car glided by and parked several slots removed from where Greg sat. A balding, middle-aged man wearing dark glasses and a suit extracted himself from the low-cut vehicle and stretched his legs. After taking an admiring look at his car and brushing a speck of dirt or dust from the front fender, the man walked purposefully into the nursing home. A moment later, Emilio drove up in his truck.

"I see Conrad decided to grace us with his presence," Emilio chuckled as he and Greg made their way into the facility.

"Conrad?"

"He's Maria's nephew. He's also the prick lawyer who's handling the logistics of the land sale." They turned down the hallway that led to Maria's room.

"Your opinion of this Conrad fellow doesn't seem too high," Greg noted. Emilio just smiled and knocked lightly on Maria's door.

"Come in," the old woman responded in a feeble voice. When they entered the room her face lit up and she struggled to her feet. "Oh my goodness, Emilio, it is so good to see you!"

"Hola, Maria, it's good to see you too." In Spanish, Emilio continued, "Maria, this is a friend of mine, Greg Martinez."

"Much gusto, Señora," Greg said with a smile.

"Oh, the pleasure is all mine. Please, won't you sit down?" Out of politeness to Conrad, she said in broken English, "And, Greg, thees ees my nephew, Conrad."

Noting the way Maria addressed Conrad, in broken English, Greg realized that her nephew didn't speak much Spanish. He followed her lead, "It's a pleasure and an honor to meet you too, Conrad," he said. Conrad responded with a disingenuous smile, his eyes conveying little more than annoyance.

From the start of their visit, the atmosphere in the room was tense. There was a subtle enmity between Emilio and Conrad that made Greg ill at ease. Emilio talked to Maria about his work on the ranch and on the movie set. He stayed away from the elephant in the room, mention of the murder of this poor woman's husband.

Occasionally he would glare at Conrad, a gesture the lawyer returned tit-for-tat. Greg's discomfort mounted. He began to wish he hadn't asked to tag along on this interview. For her part, Maria seemed oblivious to the animosity between Emilio and her nephew. In fact, she reveled in having so much company. She took an instant liking to Greg and listened in rapture as he told her about himself, his family and his father's farm. She responded with anecdotes of her own about growing up on a ranch and how much she and Augusto enjoyed visits from Emilio and from Conrad when he was a child. As long as the subject was light, Maria basked in the social atmosphere, but when Greg finally mentioned the land survey and the business nature of their visit, and in particular when he began asking questions about the deed history, she became much more reserved.

"This may not be a very good time to bring all of this up," Greg said, trying to wade into the subject with tact, "but it may help to sort out what took place on your ranch last week. You may remember my friends Rosa and Boniface Romero, the couple you spoke with the other day. Well, they feel that the 'old papers' you mentioned might have had something to do with Augusto's death." He turned to Conrad, "Conrad, I understand you've been involved in the land sale to Santa Teresa pueblo. Have you seen the old documents Maria has been referring to? It might really help . . . ."

"You've got some nerve," Conrad snapped. "This isn't the time or place to be asking discomforting questions of this poor woman." He leaned forward menacingly, "My aunt has been through a lot recently. You two shouldn't be dropping in unannounced to rub salt into fresh wounds. I don't mind it if you pay her a social call, but I won't allow you to upset her." He sat back, "Besides, the only people who have a right to snoop around in her business are the authorities. I should sue your asses for aggravated assault." He adopted a supercilious air and turned to Maria, "Disculpe me, Tia," (please excuse me, Auntie) he said in halting Spanish, "I'm just concerned for your well-being . . . pienso de Usted."

Greg was stunned by Conrad's outburst and also a bit chagrined. The guy did have a point. Less than a week ago Maria's husband had been brutally murdered. She had to be hurting still, and he'd lost sight of that in his zeal to help Denny. He now wished even more fervently that he hadn't decided to come along to bother this poor elderly woman.

"Esta bien," Maria said lovingly to her nephew. She then spoke to Emilio and Greg. Greg noticed a significant change in her demeanor. She was nervous but not upset. "I want to help your friend, Greg, and I want to find out who would think to harm my wonderful Augusto." She glanced at Conrad and then looked down. When she continued, she did so quietly, almost in a whisper, "Augusto kept a copy of the papers at the house. That is all I can tell you."

An uncomfortable silence hung over the room. Finally, Emilio spoke up, "I'm sorry to have bothered you, Maria. If it's all right, I'd like to look for those documents. Like Greg said, they may shed some light on this whole sad affair. I've got that spare key to your house; would you mind if I have a look this evening?" Maria smiled and nodded. "And Greg, you should probably come with me. I got no idea what it is we're searching for. As far as time goes, I got chores to get done 'fore Lily gets back from her class. She ought to show up around 4:30 or so. I also got to call Bobby about this. He trusts me, and because all the papers we're looking for will probably be in Spanish, I don't think he'll mind a little help with his investigation. But he won't like it none if we don't ask permission first. What say we meet at the Benavides house at about 5:00?"

Greg agreed willingly. The friction in the room had reached the point where any excuse to leave was a godsend. Besides, he now had what he'd come here to find out. They had a lead to the location of the documents which might help explain why Augusto had been killed. He and Emilio said their "goodbyes" and left Maria and Conrad to communicate in the "Spanglish" dialect they'd grown accustomed to.

AT THE FAR END of the hallway, the door opened with a metallic click. Wayne entered and walked toward Denny's cell carrying a familiar-looking satchel. It was the one that contained his clothing and personal effects. The smile on the guard's face was genuine as he unlocked Denny's cell door and held it wide open. All of the employees at the jail liked Denny. He wasn't a typical prisoner, and none of them imagined for a moment that he could be guilty of killing someone in cold blood. Wayne felt privileged to be the one to liberate him from his confinement.

"Mr. Carswell, sir, you are free to go—Sheriff just called and gave me the good news." He handed Denny his stuff, "and you'll never guess in a million years who they've arrested for Mr. Benavides's murder."

Denny dove into the canvas bag and began laying out his street clothes. He was more than ready to leave this place. As he fastened his watch around his wrist, he noted the time, "I suspect Bella Starr has been charged with the crime. Furthermore, she's probably in custody and en route from Mora as we speak. The killing took place here in Taos County. The sheriff wants to hold her here." Wayne's jaw dropped. It wasn't ten minutes ago that he'd learned of Bella's arrest. How the heck could this prisoner know so much? "Has there been any word from Paul and Camille?" Denny asked as he tightened his belt and began donning his socks. "Oh, and has Sheriff Wade returned from Abiquiu yet?"

Wayne couldn't believe what he was hearing. The only place he knew of where the prisoners knew more than the guards was at Stalag Thirteen, the POW camp in the TV show 'Hogan's Heroes.' He suddenly felt like he was the "Sergeant Schultz" of the Taos County jail—'I know nothing!' Come to think of it, Sheriff Wade reminded him a bit of Colonel Klink.

Denny handed his orange jump suit to the guard and patted him on the shoulder. "It was great getting to know you, Wayne, despite the circumstances."

Wayne finally found his voice. "I, uh, asked Jerry over at impound to bring your truck around front—keys are in the ignition."

"Thanks, Wayne." It would have been polite to chew the fat for a while with his former captor, but Denny just didn't feel up to it. He wanted out. "Hey, I gotta go and take care of some business in Arenas Rojas. You take care." The two of them shook hands and Denny walked briskly down the corridor to the front of the jailhouse. The last time he'd come this way he'd been shackled and under guard, subdued, humiliated and fearing for his life. He was now a *free man!*

As he strode out the front door into a cool, blustery breeze, his spirits soared. God it felt good to be outdoors again, to wear real clothes and to feel the wind blow through his hair (what little he had left). His incarceration had lasted less than six days, but it seemed like an eternity. Never again would Dennis Carswell take his freedom for granted. He stopped in the parking lot and looked around for his truck. He spotted it parked off to his right between two patrol cars. Suddenly his emotions began to catch up with him and he began to feel light-headed. He noticed an aged wooden bench near the sidewalk, situated in a grove of spindly Russian Olives. He decided he needed to sit for a moment to settle his nerves. A young woman stood nearby smoking a cigarette. They exchanged smiles and she took a few more puffs before grinding the butt under the toe of her shoe. Break time was over. She sighed and wandered away, heading slowly back to her daily grind. Denny almost spoke up to remind her to appreciate her job and her freedom. He didn't, of course. If someone had said that to him two weeks earlier, he'd have hurried away and not looked back. Only crazy people said stuff like that.

Denny thought back to when Roy had asked to smoke a cigarette the day after he'd landed in jail. Although Denny had a visceral aversion to the disgusting habit, he nonetheless resented the fact that Roy had been refused that simple request. When he himself asked for a calculator or a table of logarithms with which to perform important computations, he had been denied as well. To have one's life so completely controlled by others, to be subjugated to live—no, make that "exist"—in a small enclosure and be fed and treated like a caged animal, was incredibly dehumanizing. It's no wonder, he realized, that the penal system did so little to prevent repeat offenders. He envisioned that if he'd been locked up for what? A year? Two years? He too would have emerged with a jaded, pessimistic attitude towards the society that had treated him so contemptibly.

Of course, most prisoners were guilty. In order to protect the public from dangerous behavior, imprisonment was an unfortunate necessity. Besides, if Denny had, in fact, committed a heinous crime, he knew his incarceration wouldn't have been so frustrating. Unlike other beasts, human beings are capable of fathoming the nature of an immoral act and can understand the logic behind the resulting consequences.

The brief reflective interlude helped immensely. Denny stood and made his way to his truck, gulping in fresh air, still enormously relieved to once again function as an individual in a free society. He flipped open his cell phone and regarded the small battery icon in the corner of the display—less than half charged. Five days of sitting around will do that, he admitted. There was still enough juice left to call Greg, though. He dialed the number and his apprentice answered right away.

"You've reached the OK corral, Wyatt Earp speaking."

"Hola, niño, que pasa?"

"Hey, boss, congratulations!"

"Congratulations?"

"Yeah, you're free aren't you? Caller id shows you're calling from your cell. I bet they don't let prisoners use personal phones."

Denny was impressed. "That's a very astute observation, youngster. There may yet be hope for you as a surveyor. Hey, I just wanted to let you know I'll be in Arenas Rojas in "poco de tiempo," an hour from now, maybe less. Let's meet at a diner in town called Alice's Restaurant. I'll buy you dinner while you fill me in on what's been happening today."

"All right! I haven't had a decent meal since breakfast."

Denny smiled and cringed. He'd have to hit a cash machine on the way out of town, knowing Greg's appetite. But the young man would have some stiff competition this evening. Denny was famished. He must have dropped at least five pounds during his short prison stint, and the thought of real food had him salivating. He couldn't wait to attack another enchilada special or a bowl of green chili stew.

"Oh, boss, I almost forgot. Let's meet later, at a little before six. Emilio and I have a date to go through Augusto Benavides's files. We're gonna see if we can dig up the old land records that Maria was talking about. We paid her a visit this afternoon." He paused, "I also got to meet her nephew. He's a real jerk if you ask me. Maria was great, though. She gave us permission to go to her house. She's a strong woman—wants to help you and find out who killed her husband."

It was obvious that Greg didn't know about Bella's arrest, and right now wasn't the time to talk about it. Sheriff Wade would fill everyone in on the details whenever he was ready. "Six it is. See you soon, Greg, or is it 'Mr. Earp'?"

"Call me Wyatt."

Denny turned off his phone. He pulled the charger out of his briefcase and hooked it up to the cigarette lighter on the dash. Without a backward glance, he threw the vehicle into gear and pulled out of the Taos County Detention Facility parking lot. It was time to push the recent past aside and look forward. There was no doubt that it was Bella who shot her neighbor; Wayne's reaction when he'd mentioned her name confirmed that. Horace Begay and Bobby Wade were now looking into the details of her relationship with Dirk and their financial situation. The monies due them from the pueblo land deal had to be the primary driving force behind Augusto's murder, but there was something else going on here as well, something uncovered in the discussion that Boniface and Rosa had had with Maria. Boniface had the uncanny ability of being able to piece together seemingly trivial and unrelated information with regard to land conveyances, survey data, physical boundary evidence, etc. in order to resolve issues involving land title and ownership. He was, quite simply, the most knowledgeable land surveyor and problem solver Denny had ever known. If Boniface suspected there was more to this whole business than that land deal, then Denny believed that as well. The musings of a nonagenarian widow, who passed her time by lighting devotional candles to her patron Saint, had given his good friend and mentor cause to search for more information. Denny knew that whenever Boniface disappeared the way he had for the past two days, he always resurfaced with something important. He couldn't wait to see what Greg and Boniface had been up to.

LIGHTS WERE ON in the Benavides house as Greg Martinez rolled his truck to a stop. It was about a quarter to five. Emilio must have arrived early. He shut off the engine and exited the vehicle. "Emilio, are you there? It's me, Greg," he announced in a loud voice. Walking slowly along the flagstone path that led to the front door, Greg repeatedly yelled, "Hello," to anyone within earshot. He had no idea if there were others besides Augusto and Maria who occupied the house, and, to anyone other than Emilio, he would be a stranger. In rural areas such as this, people might not appreciate unexpected guests at this hour. He didn't want to be shot as an intruder. Continuing to announce his presence, he cautiously pushed open the door. Whoever was there had left it partially open. His senses went on high alert. It was cold out, and windy; something wasn't right.

"Anybody home?" He had to push firmly in order to squeeze through the doorway. Something was blocking the door. He peered inside and saw that it was an overturned chair. Beyond the chair were coats, hats and gloves scattered around, along with boxes that had presumably been stored in a closet by the entryway. The living room was pure chaos. Books from broken bookshelves littered the floor. All of the cushions had been removed from couches and chairs, most of which had been torn or cut open. Pictures and paintings had been ripped off the walls and slashed. When he looked down the short hallway leading to the kitchen, he could tell the situation there was similar. Silverware was strewn across the floor along with pots, pans and broken dishes.

Greg carefully backed out onto the porch. Vandals had been here. The place was a crime scene and he knew he shouldn't touch anything before the police arrived. He remembered reading somewhere about clever crooks who monitored the newspaper to gather tips about what houses to burgle. If that was the case, there might as well have been a neon sign out along the highway with an arrow pointing to the Benavides house:

'Rob this place. Husband dead. Wife in nursing home. Nearest neighbors more than a mile away.'

He began walking back to his truck. Knowing Sheriff Wade, and how he'd treated his innocent boss, Greg began to worry that Wade would wrongly conclude that he'd had something to do with this. Were his fingerprints on any of the stuff? He tried to remember exactly what he'd touched. Fortunately, the door had been open and he hadn't touched the knob. During the brief period he was inside, all he'd done was to look around. Greg breathed a little easier and pulled his cell out of his pocket. He would call Denny and let the old man be the one to call the sheriff. Standing in the driveway, he hit Denny's number on speed dial. After four rings, the call went to voicemail.

"Hello, boss, this is Wyatt Earp. Hey, I'm out at the Benavides place waiting for Emilio. We were gonna look for those papers of Augusto's. Anyway, Emilio hasn't shown up yet, but someone else was sure here. The place is trashed; I mean there's stuff everywhere . . . ." While Greg relayed his description of the state of affairs, two headlights signaled the approach of a vehicle. "Well, it looks like Emilio's drivin' up, I'll call you later when . . . I'll be damned, now that's a heck of a truck. It's an old sucker . . . ."

He never finished his sentence. Fireworks suddenly went off in Greg Martinez's skull. The phone fell from his hand and he crumpled to the ground, unconscious, bleeding from the blow he'd taken to the side of his head.

THE FAINT TWANG of a Country tune drifted out of Alice's Restaurant and into the parking lot. Denny noted quite a few more cars parked outside than had been there a week earlier. Business was good. Inside, the music mixed with the laughter and conversation of patrons on this Friday night, the beginning of the weekend. A typical small-town mélange was present. A family with two small children sat at a far booth. On either side of them were older couples, and at the far end of the counter a young couple gazed into one another's eyes while eating sundaes of monstrous proportions. Denny made a mental note to tell Greg to save plenty of room for dessert.

Jenny recognized him at once and greeted him with a smile. "Howdy, Mr. Surveyor, you still workin' up at the Stetson place?"

Now *that* is a loaded question, he thought with a chuckle. Rather than give her an earful, Denny simply replied, "Just about finished."

"Did you get to talk much with Dirk and Bella?"

"Why, yes . . . yes I did."

"It used to be quite the deal here you know, having movie stars livin' nearby." As usual, Jenny was a blur of motion, busing and setting a table as she talked. Although her back was now to him, she conversed as if they were face-to-face. "Now all the locals just treat 'em like regular folk; we're used to having them around. Dirk doesn't come into town too often, but Bella does. She's real friendly. Most of the ladies think she's maybe a bit too friendly, if you know what I mean. Some of 'em keep a close eye on their menfolk when they see her on the street." She turned and brushed an errant lock of hair out of her eyes, "So, you want to sit at the counter?"

"Not this evening," he replied, "I'm expecting a friend to show up soon. A booth would be better."

"Then y'all can take this one right here. I just cleaned it. Coffee?"

"Yes, please." *Real coffee,* Denny mused as Jenny floated over to the booth where the family sat. *What a treat, something to savor until Greg gets here.* He heaved a sigh of contentment and kept an eye on the platters coming from the kitchen. A benefit of his early arrival was that he could leisurely window shop for what he would order when the time came. Yes indeed, things were really looking up. He was once again a free man, his name and reputation had been cleared, and a hot meal and good company were in his near future.

During this peaceful, reflective interlude, an emergency vehicle, siren blaring, swept past the diner headed for the highway. Many of the customers stared out the window after it. Such an occurrence was somewhat of a novelty in a town this size. Denny's attention was drawn to Jenny's emergence from behind the counter bearing a large pot of coffee. He righted the upside down cup at his place setting in preparation for the moment when she would dispense the magic elixir he so craved.

"Looks like Ernie and Bert are headed toward the Stetson ranch." The words were uttered by an elderly patron seated near the front door.

Denny perked up as Jenny approached. "Ernie and Bert?"

"Those are our two EMTs," she replied, "Ernestine Lovato and Charlie Burt. People here refer to them as Bert and Ernie, like on 'Sesame Street.'"

"Yep, I see 'em turning into 'ole Dirk's driveway. That's where they're goin' all right." The old man seemed impressed with his powers of observation.

"Excuse me," Denny said as a feeling of unease came over him. He threw a couple of dollars onto the table and told Jenny he'd be back. Walking out to his truck, he disconnected his phone from the charger and turned it on. When the unit acquired signal, an icon appeared indicating that he had two unheard messages. One was from Pamela, his wife, whom he suddenly realized he hadn't spoken with in almost a week. That was sure going to cause him trouble. The other was from Greg. Denny's sixth sense told him something was wrong. He hurriedly accessed the message from Greg. As he listened, his feeling of unease turned to dread. Hitting the "save" button, he replayed the message again, fighting the impulse to panic, trying to remain calm and analytical. The only way he could be of help to Greg was to remain stoic and respond appropriately. Having almost memorized the recording, he snapped the phone shut. Now what? He had to think. His first impulse was to start his truck and drive like a madman—but to where? To the Stetson place, follow the ambulance? Or to the Benavides house?

Denny decided he needed more information. The satellite sheriff's office in Arenas Rojas was only a block away. Pocketing his cell phone, he shut the car door and jogged to the office. Camille Baldonado was behind the desk. Her face lit up in recognition as he approached her. "Why, Mr. Carswell, welcome back . . . ."

Denny interrupted her, "Camille, I wish we could chat, but this isn't a social call. Something's wrong. That ambulance—what's going on?"

"Sir, it's police business, if you want to talk to Sheriff Wade . . . ."

"Where is he, Camille? The Sheriff, did he . . . was the ambulance call to the Stetson place?" He was exasperated. "I have a friend who may be in grave danger, and I think I can be of some help to Sheriff Wade. Please, Camille, what's happened?"

The concern in his eyes won her over. "You're right. The Sheriff, Deputy Viera, the ambulance, they're all up at the Stetson house. There's been an accident. A woman called it in; she was really upset."

Denny thanked her and ran back to his truck. "There hasn't been an 'accident,'" he growled to himself as he fired up the engine. "It's this damned land transfer again, and I haven't a fucking clue what's going on." He spun out of his parking spot and churned up gravel until the car hit pavement. All of the stress from his imprisonment welled-up in him again as he fought hard to maintain his composure. Denny's foot pressed the accelerator hard to the floor until he screeched to a halt at the towering gate at the entrance to the cursed ranch of the aging film cowboy, Dirk Stetson.

XIV

Strong hands supported a groggy Greg Martinez. He was vaguely aware of being pulled from a car, ushered into a dimly-lit room and being assisted into a bed. Everything was spinning around him as though he were on some sort of amusement ride, his brain trying, unsuccessfully, to stabilize his surroundings. He closed his eyes and drifted off only to be awakened by the sensation of something cool being pressed across his forehead.

"Don't go to sleep on me, man. Doctors say that's bad karma."

The cold compress felt good, and soon the room began to slow down. As focus returned, Greg beheld a strange sight. He was engulfed in a sea of electronic equipment, the majority of which was in some form of disrepair. Most of the items were old computers and peripherals, but there were also radios, television sets, even an old toaster littering several small tables and crammed onto shelves. Above him, peering down, was a man silhouetted by an overhead light. He was young, with streaks of blonde hair protruding from a head covered with tin foil. Greg freaked. "What the . . . ?" He tried mightily to get up, but as he did a wave of nausea swept over him and he collapsed back onto the bed. More cold compresses followed. Again the spaceman spoke, "You took a heavy shot, amigo. That dude layed you out like a sack of cement."

Greg lay still for a moment while consciousness made a slow return. His head hurt like a son-of-a-bitch, but as long as he didn't open his eyes too wide, or move his head too fast, he felt better. "You mean you didn't sandbag me?"

"Me? No way, man. It was another dude. He clobbered you just as I drove up. You're lucky I came by."

"Did you . . . did you see him, recognize him?"

"Nope. It was too dark out. He ran off when he saw my truck. My name's Xeno, by the way."

Greg attempted to smile, "Hey, Xeno; I'm Greg." He opened his eyes and tried to raise himself to his elbows, but winced in pain and fell back.

"Gregory, my man, you have what they call a concussion. I was gonna drive you to the hospital, but you started to thrash around a bit. I didn't know if I could take my eyes off you in order to drive; head wounds are tricky."

"So, where are we? What's all this stuff?"

"Welcome to 'chez Xeno,' my friend. And this 'stuff' represents the crude state-of-the-art integrated circuitry of our 'modern society' being put to good use by yours truly in a valiant attempt to contact and befriend the extraterrestrials who've been monitoring us for years."

"Oh, great," Greg thought, "I've been rescued by a nut job." He closed his eyes, "Maybe it's all just a bad dream. If I keep quiet it should all disappear . . . ."

"Hey, Greg, dude, don't you be falling asleep on me now. When I think you're well enough to travel, I'm runnin' you straight to the hospital. Concussions can seriously fuck up your day, man. Nasty shit happens if you get swelling around your brain."

"Can't you just call for an ambulance on your cell phone?"

Xeno smiled and shook his head, "Too dangerous. Even with this Faraday cage," he patted the aluminum foil carefully molded around his cranium, "I can't risk it. People don't realize when they use those phones that they're holding powerful microwave transmitters right up against their grey matter. You just relax, Gregaroo, while I download some of last night's data."

Xeno briefly left the cabin and returned carrying an old laptop. He sat at the table next to Greg and carefully cleared a spot to set up the computer and mouse pad. Greg watched in fascination as his host stared intently at rows of numbers and pictures that flashed across the screen. Despite the pain in his head, and occasional bouts of nausea, Greg sat up and tried to make sense of what his weird comrade was doing. After a while Xeno looked at him and smiled.

"They come down so fast you can hardly see 'em. Either that or they just beam up their loot. See that?" He pointed to something on the computer screen, "it might just be an artifact, but . . ." he leaned close to the image and scrutinized it, "it could very well be moonlight shining off of a transport vehicle. Let's see if we can enhance the image." He tapped on the keyboard and enlarged part of what appeared to be an infrared photo of a field and some trees. "Ah ha," Xeno muttered in triumph, "I'd call that an event!" All Greg saw was a white spot. "It looks like my problem now lies in the limited resolution of the camera. I need more pixels," he lamented. "When I zoom in I just go from small and fuzzy to bigger and fuzzier."

Greg was completely miffed. "What the heck are you looking at?"

Xeno turned and smiled at him. "'Gas Buggy,' Amigo." He turned his chair around to face Greg. Sitting back, he adopted the tone of someone narrating a documentary. "The date: December tenth, nineteen sixty-seven. The time: 19:30:00 GMT. The place: latitude 36.68 north, longitude 107.21 west, near the city of Farmington, New Mexico. What took place back then shook the ground for miles around with twenty-nine kilotons of nuclear fury." He stared at Greg for signs of comprehension, but all he got in return was a puzzled look.

"It was an underground nuclear test, Gregaroony, an event named 'Gas Buggy,' one of only two shots conducted in the great State of New Mexico. The other one, of course, was 'Trinity.' That one took place way down south, outside of Alamogordo." He leaned forward, his eyes bright, "Can you now see what they're after?" Greg still wore a look of total bewilderment. "They have to refuel their ships, man, don't you get it? They want plutonium, enriched uranium, whatever remnants of fissile material are down there. If one wants evidence of intelligent extraterrestrial life, what better place to stake out than a filling station? When people travel by car, they stop at gas stations. When space travelers are running on empty, they go to places like the 'Gas Buggy' site. I go once or twice a month, whenever there's enough moonlight to see 'em."

Whatever Xeno was talking about he was sure into it in a big way, Greg realized. He lay back again and gingerly rubbed his throbbing skull. "I know I probably owe my life to this guy," he thought, "and he seems pleasant enough, but . . . he's nuts."

TWO POLICE CARS and an ambulance were parked near the barn at the Stetson place. Denny skidded to a halt, got out, and ran towards the assemblage of people he could see silhouetted by the lights in the tack room. A profane male voice could be heard above the idling engine of the ambulance. Denny recognized it immediately. It was that of Sheriff Wade.

"For Christsake, Ernie, can't you give me more than that? I want answers, and I need 'em yesterday, you got that?"

"That's not my job, sheriff," Denny heard the female EMT say, "You're going to have to wait for the forensics guys. I'm sorry."

Denny worked his way into the room and into the corner where the sheriff stood. What he saw on the floor took his breath away. Emilio Sisneros lay there, his head plastered with congealed blood. He was dead. Near the door that led to the stables, Lily sat with her head in her hands. Her eyes were red and puffy and her normally soft, expressive features were contorted in anguish. She'd been crying for some time. Everyone looked up as he entered the room. The Sheriff's countenance turned dark, almost hateful, when he saw who it was.

"What the hell are you doing here, Carswell? You're a free man now, but you won't be for long if you don't get out. This here's a crime scene, police business."

Sheriff Wade was wound tighter than a hangman's noose. Denny knew he had been good friends with Emilio, rodeo buddies, and this was apparently how the man handled his grief—by flaunting his authority and throwing a tantrum. It was a bad situation for all concerned. For Greg's sake, Denny had to get the Sheriff to calm down. Before he attempted to do that, however, Denny had to figure out just what the heck was going on.

Right now Denny had to brace himself for verbal attack. Bobby Wade wasn't through with him. The man's eyes were glazed over and he needed a whipping boy. His vitriol was directed straight at Denny, and the man was losing control. "What I'm gonna do is throw your ass back in jail. This was a peaceful place 'afore you showed up. Hell, you and this woman (he gestured toward Lily) are both under arrest under suspicion of murder. At least you won't be able to cause any more trouble 'till I get some real answers and can figure out what's goin' on."

The Sheriff paused and his gaze shifted abruptly as the tall figure of Dirk Stetson appeared in the outer doorway. He looked nothing like the rugged cowboy of his movies. Dressed in jeans, loafers and a sleeveless padded vest, he simply gave the appearance of a tired, haggard old man. There were deep circles under his eyes, and Denny could smell whiskey on his breath. He surmised that Dirk was in shock after hearing the news about Bella. Having the police pay him a business call was adding to his apprehension. To make things worse, his timing couldn't have been less perfect.

Wade pointed a menacing finger his way. "Why, if'n it isn't old Dirk Stetson himself come to privilege us with his presence. Tell me, Dirk, where were *you* about an hour ago? Know what? I think odds are you did this. I bet you and that skank wife 'o yours were in on this together. I get it now; she whacks the neighbor and you kill poor Emilio here. What was he, a witness to some crime you two cooked up? He was the most decent man alive, and some coward bashed his head in from behind." The sheriff was working himself into a real lather now. Denny looked over at the Deputy, Paul Viera, and saw grave concern on the young officer's face. The already strained situation was deteriorating quickly.

As if in answer to the prayers of everyone in the room, headlights and the sound of crunching gravel divulged the arrival of a vehicle. It was a modified SUV with a flashing light rack on top and on its side it bore the State Police logo. "Thank God," Denny muttered when he saw who it was. Horace Begay stepped out of the car and quickly strode into the Stetson barn.

All in attendance breathed a sigh of relief at the sight of the new arrival, even Sheriff Wade. The presence of another authority figure, especially the calm, capable Horace Begay, helped to defuse some of the tension in the room. Horace and Denny exchanged knowing looks. Denny shook his head slightly to indicate the gravity of the circumstances.

"Camille told me to hurry out here," Horace said. He looked down at Emilio and his shoulders slumped. There was nothing more he could say.

After this brief respite, Bobby Wade once again began to hurl threats and accusations. Everyone in the barn, it seemed, had had both opportunity and motive to kill his friend. During the renewed tirade, Denny cautiously moved back away from the Sheriff's view, to a corner of the room where he could think. Greg's fate was in his hands and his alone. He, Denny Carswell, was the only one present who knew that Emilio was supposed to meet with Greg that evening. The two events, their meeting and the killing, were obviously related, but how? Denny glanced at a nearby table upon which sat a folder. It was labeled 'Conestoga Highway,' and contained a loose assortment of papers: receipts of various sorts for feed and supplies, services rendered, etc., as well as filming schedules and cost estimates. It represented the business side of what Emilio did for the movie. Denny imagined how much a man like Emilio probably detested paperwork. They were a lot alike in that regard. Emilio loved working with horses as much as Denny liked his fieldwork, but the downside to any job was the drudgery of handling the necessary documentation. "Documentation," he thought. All of a sudden facts started falling into place. Greg and Emilio were to go searching for old documents that had belonged to Augusto Benavides. Earlier that day they'd met with Maria. *That's it!* he realized. The meeting with Maria, the ransacked house, the abruptly terminated voicemail from Greg . . . he knew exactly what had happened . . . and what was probably happening now.

"Sheriff," he exclaimed, "Sheriff, listen, I know . . . ." But the Sheriff merely stared him down and kept talking. Denny instantly grew furious. Lives were at stake, and he had no choice but to take charge. "Sheriff," he yelled at the top of his voice, "Shut Up!"

Silence immediately befell the room as all eyes set upon him. Bobby Wade stood in shock, surprised that Denny, or anyone for that matter, would dare speak to him like that. Horace's eyes narrowed and he became keenly alert. There was no telling how Sheriff Wade might respond. Others in the room cringed in anticipation of the shit storm that was to follow.

Denny didn't give anyone time to react. He looked the Sheriff in the eye, "I know who did this." He paused to allow the words to sink in and for Bobby Wade's temperature to drop by a few hundred degrees. He then calmly related the facts to all those assembled and used them to clearly describe what was happening. When he was done, he fumbled in his coat pocket and extracted his cell phone. He punched a few buttons and handed the unit to Sheriff Wade. The chirping of a cricket would have sounded like thunder in that room while all parties held their breath and Wade listened to Greg's voicemail message.

When he was done, Wade tossed the phone back to Denny. "Bullshit," he said, "that call don't prove . . . ."

Denny interrupted, his tone defiant and firm, "Greg and Xeno are in danger over at the Benavides ranch. If you don't settle down and get over there, pronto, you'll have two more killings to deal with, and their blood will be on *your* hands." There was, of course, the possibility that the two young men were already dead, but Denny didn't think so. There was no way he could bring himself to even think along those lines anyway. He had to maintain hope.

Before the Sheriff could respond, Lily jumped up and grabbed a bridle from the wall. She then hefted a saddle and blanket from a nearby table and hauled them over to one of the nearby stalls. Its occupant was a large stallion whose name was, fittingly, "Stud."

"Just what the hell are you doing?" the Sheriff asked in an ominous tone.

Lily didn't answer. Quickly and efficiently she prepped the animal and moved to place her foot in the stirrup. Only when she heard the characteristic click of a pistol hammer being cocked, did she turn to face the Sheriff. His gun was drawn and he was pointing it at her horse. "You seem to forget, young lady, that you are a murder suspect. You're not going anywhere."

"I'm going to do what Mr. Carswell says," she said firmly.

Horace interjected, "I think Carswell's right, Bobby. I trust his judgment. Remember, he figured out that Bella killed Augusto, and I've worked with him in the past. Hell, I've never known him to be wrong."

Lily spoke again, "We both loved Emilio, sheriff. I, for one, want to catch his killer, not stand around here playing macho games. And, sheriff," she added with conviction, "I could really use some help."

The Sheriff stared at her and then at Begay. Lowering the hammer on his revolver, he returned it to its holster. "Damnit," he muttered, "she's right." He turned to Paul Viera, "Stetson doesn't leave your sight, got it?"

Lily ran and grabbed another saddle. "It's faster if we ride, sheriff."

"Now, this is more like it," Denny thought. He caught Lily's attention as she and Bobby started to leave. "You two swing by Xeno's cabin. We'll rendezvous at the Benavides house."

"I'll drive," said Horace.

"Wait, Mr. Carswell," Paul intervened, "you can't go anywhere."

"I'm not a prisoner anymore, Paul. I'm going to help my friend."

Paul considered Denny's words and nodded. He then turned to Dirk. "Dirk, get in the patrol car."

"But the Sheriff said I have to stay here."

"No, what he said was that I'm not to let you out of my sight, so get in the car, they may need backup."

"AH, THERE YOU ARE, both of you." Conrad Sussman stood in the doorway of Xeno's shack. There was an evil grin on his face and he was pointing a gun at the others. "Let's see, what have we here?" he clucked while he looked around the small structure, examining the enormous quantity of used equipment. He soon spied something of interest. Shoving a box of tools off a bench, he grabbed a bundle of cable ties. "Ah, perfect. Xeno, stay in that chair and put your hands behind your back. If you try to play hero, well . . ." he waved his pistol menacingly, "I would strongly advise you not to." He bound Xeno's hands both to one another and to the chair before rudely shoving Greg onto his side and securing his hands as well. When he was satisfied that his captives posed him no threat, he stuck the pistol in his belt and sat side-saddle on the table beside Xeno's computer.

"Where is it?" he asked matter-of-factly.

"Where's what, dude?"

"Wrong answer," Conrad snarled, and delivered a savage back-handed blow to Xeno's chin. "Son-of-a-bitch," he yelped, grabbing his knuckles. "Geez you got a hard fucking head." He arose and began pacing around the room, shaking the sting out of his hand. "Man, what a shit hole. The only reason why any self-respecting person would want to live in a dump like this would be 'cause he's trying to get his hands on something—something of Augusto and Maria's. Actually, it belongs exclusively to Maria, but when she dies, *heh, heh*, it will belong to me."

Greg and Xeno exchanged baffled looks. Neither had any idea what the lawyer was referring to.

Conrad scowled. He traipsed back and forth from one end of the cabin to the other, pulling items off of shelves and knocking things off tables and benches. Occasionally he would throw an object onto the floor and scavenge through the resulting pieces. As his search continued to prove unproductive, Conrad became more and more agitated. Greg recognized the handiwork. "You're the one who ransacked the Benavides ranch house, aren't you, Conrad?"

"Brilliant deduction, Sherlock. Of course it was me; I tried to find it before you got there. When you finally showed up, I watched to see where you went to look, but all you did was stand around with your head up your ass and then leave." He paused to reflect on what he'd just said. "That means one or both of you must have brought it here."

"Brought what here?"

"The land grant, you moron."

"I still don't know what you're talking about," Greg said in a tired voice.

"We've got all night, my friends, and believe me, I can make life very unpleasant for both of you if I choose to. Sooner or later you're going to tell me what I want to know, and once you do," he shrugged, "why, then you'll be free to go."

LILY LED THE WAY along the normally easy-to-follow trail as she and Bobby gradually made their way towards the eastern portion of the Benavides ranch. Despite the fact that it was still twilight, and a bright, yellow, almost-full moon was rising in the east, visibility was poor. The dense canopy of pine and fir seemed to close in on them as they moved along, branches occasionally appearing out of nowhere to whip them in the face. They were moving as fast as practicable under the circumstances. Wade marveled at how well his companion handled her mount, such a small woman exercising such complete control over such a strong, massive animal.

When they finally reached a clearing along the top of a ridge, at the same spot where Denny, barely a week earlier had witnessed Augusto's murder through the scope of his total station, they spotted a small cabin with its interior lights on. "That's where we're headed," Lily said. "That's Xeno's place. The main ranch house is about 150 yards off to the southwest." She picked up the pace, anxious to get there, sensing that time was of the essence. Once again the trail plunged into the woods, but here the trees were larger and well separated. The ground was rocky, inhibiting undergrowth, and enough moonlight got through that they could easily see where they were going.

CONRAD BIT HIS LIP in contemplation of his next move. Reaching into his jacket pocket, he extracted his bottle of nasal spray and took a hit in each nostril. He sniffed to clear his head and, with renewed vigor, made his way to Greg's bedside. Without warning, he grabbed Greg by the hair and jerked his head around. Stabbing pain shot through Greg's head and neck and he cried out in agony. His tormentor moved in close until their noses almost touched. "You're the one who knows, aren't you?" he sneered. "If Mr. Spaceman had found it during the year he was 'helping' Augusto, he never would have stayed around in this place."

"Leave him alone, Conrad," Xeno shouted. "You hit him real hard back at the house, man. He's hurt bad, and if you treat him like that he's gonna pass out. He won't be of any use to you unconscious."

Conrad thought about what Xeno said. He had to admit the "freak" had a good point. With a reluctant grunt, he let go of Greg's hair and turned to his other captive. "Okay, smart ass, let's find out how much *you* know." He extracted the sleek Ruger 9 mm from his belt and held it menacingly in the air. His face contorted into a sadistic smirk as he pressed the barrel firmly against Xeno's temple. An eerie silence ensued, broken only by Conrad's wheezy breathing. He drew back the gun's hammer, preparing to end Xeno's life in an awesome display of gore and power. If that didn't loosen Greg's tongue, nothing would.

Before he could pull the trigger, however, he felt the ground start to vibrate. The sensation increased and, suddenly, he heard them as well—horses hooves pounding on the ground nearby. Conrad jumped to his feet and stumbled to the door. He opened it part way and peered outside. In a moment he saw them, two riders emerging from the forest. He crouched low and moved to a better vantage point behind a pile of firewood stacked neatly under the eve of the cabin.

As they approached the clearing that surrounded Xeno's cabin, Lily brought "Stud" to a walk. There was no telling what might await them. She was in the lead. Bobby Wade rode a step behind and to her right. When the Sheriff spurred his mount to pull up alongside, he caught sight of movement in the sensitive periphery of his field of view. A second look confirmed the presence of a figure hiding in shadow. His adrenaline surged when he noticed a shaft of moonlight reflecting off the blued barrel of a gun.

"Yee Haw!" Wade shouted. He urged his horse forward and turned to cut between Lily and their assailant. It was a display of horsemanship he hadn't executed since his rodeo days, and for an instant his mind flashed back to some of the best times of his life, back to team roping events that he and his partner, his good friend Emilio Sisneros, would enter in weekend competitions all over New Mexico and west Texas.

The crack of the weapon shattered the stillness of the forest, and its echoes reverberated off the walls of the cabin and through the surrounding hills. Although the shooter's technique was poor, he got lucky. The bullet struck the Sheriff in the upper right arm and passed through his shoulder. The impact knocked him from his horse. Lily jumped off of Stud and ran to where Wade had fallen. Conrad fired twice more, but darkness and a shaky hand resulted in wild misses. Grabbing the sheriff's good arm, Lily helped him stagger over to a nearby tree.

Blood oozed through his shirt and a growing stain formed on his jacket. In twilight it was hard for Lily to assess the severity of the wound, but the limited bleeding indicated that no artery had been hit. The sheriff also exhibited no difficulty breathing, which suggested that his lungs and heart were unscathed. All things considered, he was lucky. After an initial course of colorful profanity, Wade's adrenaline level dropped and he relaxed against the trunk of their sheltering fir tree.

Bobby reached across his body and drew his revolver from its holster. Unfortunately, he was right handed, and, given the trauma he'd just suffered, he knew he couldn't effectively use the weapon.

"Here, take this," he half-groaned, "I want you to shoot that sum-bitch!"

Lily took the gun, but set it down in order to tend to the Sheriff's wound. Conrad fired again, and this time the bullet nearly found its mark, embedding in the fir tree with an audible thud.

"Dang it," Wade said, "Stop fussing with me. I'll be all right."

Lily nodded in grim assent. The Sheriff needed professional medical help. To be of any real assistance she had to get to a phone. Perhaps the most logical course of action would be to make her way to the Benavides house where the State cop could radio for help. But the Sheriff might need her here, and there was, of course, another compelling reason to stay: Xeno. No, make that two reasons: Xeno and the young surveyor, Greg Martinez. She had to find out if they were okay.

She stood and peered around the tree. A flash of fabric gave away Conrad's position. He was behind a wood pile just to the left of the cabin door. Holding the gun with both hands, she aimed and pulled the trigger. The resulting thunderous boom took her totally by surprise. She'd fired hand guns before, but never anything like this. The 300 grain slug left the barrel travelling almost 2000 feet per second and slammed into the wood pile. The effect was devastating. Logs and splinters flew back and slammed into the cabin, dissipating the more than 2000 foot-pounds of energy imparted to them. It was as though a giant, wielding a sledge hammer, were venting his frustrations on a pile of sticks.

The effect on Lily's end was just as impressive. To a student of physics it was a beautiful illustration of Newton's third law of action and reaction. She hung on tight as the gun kicked back, lifting her completely off the ground. She landed ignominiously on her derriere in a small patch of scrub oak. "Son of a bitch," she muttered as she crawled back behind the tree and got up on one knee. She examined the top of the barrel in the light of the nearly full moon which now sat majestically above the trees:

'S&W .500 mag.'.

"Hell, this is no pistol; it's a freakin' canon!" She cast a contemptuous glance at the Sheriff. "I should have figured as much," she mumbled, "a big macho gun for a big macho prick." She immediately chastised herself, however, for the disparaging thought. Moments earlier, that "macho prick" had selflessly and heroically saved her life. As she looked at him, she became more concerned. He looked pale, an effect, she could tell, from loss of blood. Lily crawled over to him and took off her jacket and belt. He regarded her wordlessly as she balled-up the coat and cinched it tightly against his arm, staunching the bleeding. She smiled at him and he nodded gratefully. Her smile, however, masked a growing fear. If he didn't get help soon, he wouldn't make it.

Lily had to do something. "Conrad," she yelled, "Conrad, give it up. Reinforcements are on the way."

After the wood pile exploded in front of him, Conrad Sussman had run back into Xeno's cabin and was now leaning against the wall, breathing heavily. At the sound of Lily's voice his fear lessened. So, that's who had shot at him. Well, if she was unhurt, who was the other intruder he'd shot off his horse?

"Well, well, if it isn't Lily. Come to rescue your man, have you? How touching."

"You shot the Sheriff, Conrad. He's injured, but he'll make it if we can get him to a hospital. You're a lawyer. You should know you can't afford to let him die. It'll mean the death penalty. Give it up."

"So," thought Conrad, "that's who I shot, Sheriff Wade." The thought emboldened him. Sheriff Wade was a real cowboy, someone who'd been around guns all his life, yet, he, Conrad Sussman, a lawyer and novice shooter, had taken the man down with one shot. A feeling of power and omnipotence came over him. He was playing for keeps now, and winning. He had to keep it up, though, every move he made from here on had life or death consequences for him. Killing Emilio had started this whole mess, and Lily was right; he would be in deep shit if he ever got caught. But for some reason the whole thing invigorated him. All of his senses were sharp, acute, operating at 110%; he felt more alive right now than he had at any other time in his life. There was only one way he could survive this ordeal now, but survive it he could, and the first order of business was to get rid of Lily. He sneered as a clever idea came to mind, a plan to accomplish two things at once. Before he killed her and the others, he'd use her to get the information he needed.

Conrad stumbled over to where Xeno sat. He cut the cable tie that anchored him to the chair, but left those that bound his wrists together. Grabbing his prisoner by the ear, he cruelly pulled him over to the open doorway, being careful to keep himself out of any line of fire. Shoving his pistol against Xeno's head, he yelled to Lily, "Let's not screw around, Lily. Here's the deal; it's simple enough. I'll count to three and then blow Xeno's brains out. If you would prefer that doesn't happen, then throw down your weapon and get your ass in here."

Lily was elated to see that Xeno was alive and well, but her gut tightened when she considered he might die within seconds, right in front of her.

"Don't do it, Lily," Xeno shouted, "he'll kill both of us. Stay put!"

She looked longingly at the handsome, quirky man with the foil-lined hat. Tears rolled down her cheeks as she thought about the kindness and support he'd shown her in the past year. She was in love with him; it was no secret. Everyone at the ranch knew it, even Conrad. That's why the creep was holding him hostage.

Conrad ground his gun against Xeno's cheek and announced, "If I were going to kill him anyway, then why isn't he already dead? Huh? Like I said, no more screwing around. The count starts now . . . one . . . ."

Lily glanced over at the Sheriff. He met her gaze and shook his head. Too weak to sit up, he said in a hoarse voice, "Just shoot the bastard."

". . . two . . . ."

She was out of time. If she tried to shoot Conrad there was a strong possibility she'd hit Xeno. That was a risk she was unwilling to take. Besides, she'd be lucky to hit the side of the cabin with the damn artillery piece that was in her hand. "All right," she said, "don't shoot. I'm coming over—see?" She held the gun over her head and stepped into the clearing.

"Ditch the gun," Conrad commanded.

Lily nonchalantly tossed the weapon off to the side, to within about thirty feet of where Bobby Wade lay. Maybe the Sheriff could summon enough energy to grab it and come after her. Another quick look in his direction, however, dashed any such hope. Wade was out of it. Blood loss had weakened him beyond the point of effectiveness. Lily took a deep breath and marched towards the cabin.

XV

Conrad backed away, still holding the gun to Xeno's head as Lily entered the room. Shoving the young man back into his chair, Conrad gestured at Lily, "Go stand behind him, and no sudden moves or I'll shoot you both."

As Lily walked by, Xeno whispered something to her. She bent down as though to hug him and heard, "your knife, my hands."

"Hey, no talking!" Conrad barked.

Lily gave him a disparaging look and moved behind Xeno. She carefully moved to a position where his torso shielded her hands from Conrad's view.

"Hey, Lily, that was a dumb thing you just did."

She looked down at Greg, saw the twinkle in his eye, and tried mightily to suppress a smile. It was a relief to see that the young surveyor was alive and in good spirits, but the dried blood on his face and weakness in his voice indicated that he was badly hurt. She turned her attention back to Xeno's hands but stopped when she noticed Conrad's gun pointing straight at her face.

"Okay, guys and girls," their captor said, ""I either get what I want or somebody dies." Once again he produced his spray bottle of decongestant. His hands shook while he administered a dose and then leaned against a bench, waiting for the medication to take effect.

All three captives soon noticed strange discomfiting quirks in Conrad's behavior. The man's pupils were like saucepans and they darted incessantly from one subject to another. His nervous ticks were becoming more pronounced. Conrad Sussman was wired, dangerously close to the edge of emotional overload. Xeno called attention to it in an attempt to divert his attention from the idea of killing Lily. "Hey, Conrad, what's in the bottle?"

"I have allergies, all right? And it's also none of your damned business."

"A little 'happy juice' perhaps? C'mon Conrad, everyone knows about your cocaine addiction. You need help, man."

"Me, ME? You're the one who needs help." Xeno's plan was backfiring; the lawyer was becoming more aggressive. "Look at you, you pathetic loser. What kind of idiot spends his time wiring together old waffle irons and microwave ovens in a ridiculous attempt to communicate with little green men from another galaxy?"

"Actually, they probably aren't from another galaxy, dude. My guess is that they're from one of the nearby G-type main sequence stars, like Alpha Centauri."

"Oh, for crying out loud," Conrad said in frustration, "Shooting you will not only be a pleasure, I'll be doing the human race a favor by removing you from the gene pool."

"I don't know about that," Lily said defensively. She put her hands on Xeno's shoulders, "He's a kind, warm-hearted person. The world needs more people like him."

"Ha, what a load of sentimental bullshit. You're wasting my time."

Lily noticed a darkening in Conrad's countenance. The danger level was growing. She couldn't put it off any longer. Her hands fell down to her belt where her small, folding utility tool sat in its holster. Until now she had resisted cutting the plastic ties that bound Xeno's hands. He'd been wiggling his fingers for the past few minutes, urging her to do it, but Lily knew better; the time wasn't right. Once his hands were free, Xeno would immediately lunge at Conrad in attempt to overpower him. Given the distance between them, it wouldn't work. Conrad was too alert and the gun was in his hand, ready to fire. Such a move on Xeno's part would be suicidal.

Now, however, the situation was critical. To conceal what she was doing with her hands, she desperately tried to engage him in further conversation. "Think again about what you're doing, Conrad. You might be able to argue self-defense in order to fight the charge of murdering Emilio; after all, there were no witnesses, but the Sheriff—you can't let him die. It'll mean the gallows for you."

Conrad grinned wickedly, "Oh, you are so naïve. Xeno shot the Sheriff. I saw everything. As a matter of fact, our 'nutty professor' here shot you two as well." He sighed in mock regret, "I finally had to kill him in a valiant but futile attempt to save you."

Lily was in trouble. The sheath around the utility tool had a Velcro fastener. The sound it made as she slowly pulled on the flap seemed almost earsplitting in the relative quiet of the cabin. There was no time to spare, she had to open the knife blade on the tool, but there was no way she could do it without being discovered! Conrad leveled his gun at her. "Say your prayers, Lily. You're first."

TWO POLICE VEHICLES, a State cruiser and one from the Taos County Sheriff's office, skidded to a stop in front of the Benavides' house. Horace Begay jumped out and ran to the driver's side of the other car. "Paul, I'm going to make my way around back. You stay out here in case I flush an undesirable." He began to leave, but stopped and returned, "Oh, I almost forgot, be attentive and try not to shoot *me*, okay, son?"

HE PADDED SILENTLY through the forest: a shadowy figure moving like a vaporous spirit. Born with uncanny vision, he could see almost as well in moonlight as he could in broad daylight, an evolutionary adaptation that made the night his true element. Good vision alone, however, wasn't enough to ensure his survival. Mother Nature had also blessed him with enhanced intelligence, extraordinary olfactory ability, and crushing jaw strength, all of which he used to significant advantage when dealing with both pursuers and prey.

Lights were on in the residence. He approached with extreme caution. It had been a while since his last visit, and there were smells, both familiar and foreign, that greeted him. Under normal circumstances he would abandon this effort and return to his other haunts, but this evening hunger drove him onward. Without making a sound, he made his way to the door jamb and peered around the corner.

He stared intently at the inhabitants while he gathered the courage to present himself. Although they were making some noise, there were no sudden movements. Feeling that everything was as it should be, he gingerly stepped out of the shadows and into the open doorway. There he stood with head lowered and salivary glands stimulated in anticipation of his treat. He uttered a soft whine to draw attention to himself—and then "smiled."

Lily froze. So, this was it. The smirk on this doped-up creep's face would be the last sight she would see before oblivion. If she tried to get away, she'd be shot anyway or, worse yet, Conrad would simply shift the gun slightly, shoot Xeno, and then gun her down before she was half way to the door. There was no way she could outrun a bullet. At least if she stood still, it might buy the man she loved a few extra minutes of life. Out of the corner of her eye she saw Xeno's shoulders tense, presumably in anticipation of the upcoming shot. But, just when she thought her world would end, she noticed what had really caught Xeno's attention. "What the heck . . . ?"

Conrad's back was to the door, but when he noticed the others looking that way he turned his head.

"Run, Henry!" Xeno shouted to his canine friend.

Framed in the cabin doorway was a strange, almost surreal sight. The vision of the drooling coyote, with bared fangs and yellow, predator eyes, was too much for Conrad's drug-addled brain to comprehend. Primal fear suddenly overwhelmed him. He stumbled backward, almost losing consciousness as blood rushed to his limbs in response to a massive cocaine-assisted adrenaline surge. "Ahhh!" he screamed. He stumbled and fell in desperate attempt to distance himself from the grotesque monster, the one his mind told him had come to rip open his throat in sanguine retribution for the crimes he'd committed. He scurried backward, half crab-walking, half crawling, until his head hit one of Xeno's junk-laden tables. As part of his initial shock began to subside, tendrils of reason began to filter up through the pea-soup fog of his terror. He still had his gun. Waving the weapon in the general direction of the ghastly image, he loosed a salvo of wild shots.

Bullets peppered the door frame, ceiling and far wall. The concussive noise from the blasts overwhelmed the senses of everyone in the room. A freight train could have rumbled through the cabin and gone unnoticed amidst the noise and chaos that had been triggered by the presence of one poor, hungry coyote.

It was just the distraction Lily needed. Ripping her multi-tool from its scabbard, she swiftly flipped out the knife blade and, in one clean motion, sliced through Xeno's bonds. While Conrad continued to fire at the hideous after-image, now present only in his mind, the powerful young man rushed towards him.

Henry, of course, had evaporated into the night long before the shooting started. No future offerings, no matter how enticing or tasty, could ever induce him to return to that place. Humans, he decided, were just too unpredictable to be around. It was now time for him to return to his new-found female companion, to enlarge their den under a tree in the far hills of the Santa Fe National Forest, and make puppies. Xeno had given Henry the gift of life the previous winter, and Henry had unwittingly repaid that debt this evening.

Alas, the cabin could have used a good cleaning. Scraps of plastic and metal covered the floor as a result of Conrad's earlier search. Xeno slipped just as he was about to grab the gun hand of his adversary. It was the housing of a toner cartridge from an old HP laser printer that slowed him down. His forward momentum, however, was enough to propel him into his foe, knocking him forward and sending the gun rattling across the floor. Both men scrambled for the weapon, viciously clawing at one another to get to it first. If he'd only grabbed Conrad and held on tightly, Xeno would easily have been able to subdue him, but, as luck would have it, his persecutor somehow managed to escape the young man's grasp, find his footing, and grab the gun as he staggered out the doorway.

Xeno and Lily both rushed through a haze of gun smoke and reached the door simultaneously. Unfortunately, they arrived too late to do anything other than watch as Conrad crashed through the forest on the other side of the parking area.

"Stay here, Xeno," Lily warned, "he's still armed." Lily cautiously jogged over to where she had tossed the sheriff's gun. Checking the cylinder, she verified that she had four rounds left. As she walked over to tend to Sheriff Wade, one thought occupied her mind. If that bastard, Conrad, were dumb enough to return, she wouldn't hesitate to blow a hole in him the size of Texas.

DENNY AND DEPUTY VIERA watched as a dark figure emerged from behind the Benavides ranch house. Horace jogged over to them with his gun drawn, wearing a concerned look. "His car's parked around back, but there doesn't appear to be anyone home." He rattled a keychain that glistened with reflected moonlight. "Keys were in the ignition. If he's hiding inside, we've got ourselves a dangerous situation. He's probably armed and he has nothing to lose. I'm going to call for more backup and . . . ."

The sound of gunfire broke him off in mid-sentence.

"Shots came from over there," Paul yelled excitedly, "down this drive. That's where Lily and the Sheriff were headed."

Horace nodded soberly. He'd also taken a good look at the weapon Bobby Wade had pulled back in the barn. "And those shots weren't fired from the Sheriff's gun. C'mon, it's faster if we go on foot." Denny followed Horace as he began to sprint down the driveway leading to Xeno's cabin.

Paul started to follow, but abruptly stopped and returned to the passenger side of his police unit. "Dirk, I want you to stay put 'til I get back, understand?"

Dirk nodded vehemently. Whatever the cops told him to do was gospel to him. Still in shock over the revelation of his wife's involvement in the Benavides killing, he had succumbed to efforts by Larry to cheer him up through the consumption of a considerable quantity of ethyl alcohol. Not accustomed to heavy drinking, Dirk was now experiencing the inevitable aftermath of such behavior. His head hurt and his stomach was trying to send him a message, by way of his esophagus, that it resented the way it had been treated. Also, the Sheriff's tirade in the barn that evening had exacerbated his problem. In his inebriated state he'd been unable to understand much of what had been said other than to realize that somehow he, Dirk Stetson, was now a suspect in two murders. How was that possible? This could ruin his life, his career. The few synapses in his brain that were functioning cogently were now urging him to cooperate fully with the police and to pray for a miracle. Deputy Paul had given him a direct order prior to running off to join the others, and he would obey that command no matter what happened. The order did not, however, prohibit him from vomiting should the need arise. Therefore, his immediate overriding concern was that, should he have to puke, he'd have to do it outside the vehicle. To do otherwise would not improve his situation.

While Dirk massaged his throbbing head and grew ever more disconsolate about the direction his life was taking, a lone figure staggered out of the woods and approached his car. Dirk recognized him immediately. "Hi, Conrad."

The lawyer had a wild look in his eye. Without saying a word he glanced nervously around and then jogged over to Dirk who was sitting in the passenger seat of the patrol car. After Paul left, Dirk had swung both legs out the open door to wait while his gut churned and prepared to launch an inevitable eruption. Some of his discomfort abated, however, at the sight of Conrad. It was nice to have a visitor, someone to talk to, something to divert his attention from his upcoming gastrointestinal revolt.

"You better be careful, Conrad. Somebody's shooting out in the woods."

"Yeah?"

"Yeah. Deputy Paul and the others went to investigate."

Conrad wasn't in a conversational mood. He brusquely grabbed Dirk by his jacket, "Get out. You're coming with me."

"I can't, Conrad; Deputy Paul told me to stay here."

"You idiot, as of right now you're what I call collateral. Let's go." He pulled hard, but Dirk stubbornly resisted.

"I told you, Conrad, I'll get in trouble if I go with you." He yanked his arm away and leaned back further into the car.

Conrad had had enough. He once again grabbed Dirk's coat with one hand and at the same time swung his other arm, the one holding the gun, in a vicious downward arc aimed at Dirk's head. As luck would have it, however, Conrad's aim was off. The butt of the gun glanced off the edge of the car's roof, lessening the force of the blow. Only part of the weapon found its mark; the barrel raked down Dirk's cheek, its front sight opening a long, painful, if only superficial cut.

"Ow!" Dirk cried. He threw his hands up in self-defense and accidentally knocked the weapon from Conrad's grasp. "Be careful, Conrad!" Realizing what had happened, he then reached down and picked up the pistol. "Conrad, you dropped your . . . ."

Conrad panicked. Dirk had the drop on him now. There was only one logical course of action; he had to get away. He screamed in frustration and took off, running back towards the safety of the forest, weaving in serpentine fashion lest his enemy decide to fire upon him.

"Wait, Conrad, you forgot your gun!" Dirk cried out. But the lawyer was already in the trees, crashing through brush. He'd made a clean escape and hadn't heard a thing.

HORACE BEGAY CAME JOGGING down the dirt road from Xeno's cabin. He needed backup and an ambulance to handle the two injured men on the premises. Sheriff Wade was in serious condition, but Horace was reasonably sure he'd pull through. The guy was tough, in good shape, and had a lot to live for. It would take more than a bullet to the shoulder to take him out. Overall, despite the fact that a killer was still at large, Horace was enormously relieved. Things could have been worse, much worse.

Horace's attention was uncharacteristically diverted by thoughts of the emergency at Xeno's cabin, so much so that he nearly panicked when he saw a tall figure appear before him frantically waving a gun. His hand tightened around the handle of his own weapon, but he relaxed a bit when he saw that it was just the actor, Stetson. "What's wrong, Dirk?" he asked as he slowed to a walk. "Is that a gun in your hand?"

Dirk nodded and explained, to the best of his impaired ability, what had happened. The disjointed narrative from the tipsy actor was confusing, but Horace understood what was truly important—Conrad was unarmed and on the lam. Horace relaxed and placed the all-important call for the ambulance. He then turned to Dirk, "How long ago did Conrad leave?"

"A minute or two before you got here, sir. You just missed him."

Begay contemplated the blood flowing from the cut on Dirk's cheek as he casually took possession of Conrad's pistol. He shook his head in wonder: *just what the hell happened here?* "Which way did he go, Dirk?"

Stetson pointed toward the trees.

"I'll be back soon, Dirk. You stay put."

Dirk nodded. He would do exactly as the officer said.

Fifteen minutes later, when Horace Begay returned escorting a manacled and thoroughly dejected Conrad Sussman, Dirk Stetson hadn't moved so much as one finger.

XVI

Ten days. By law, ten days was the maximum length of time a suspect could be held prior to his preliminary hearing. The prosecutor had asked for, and received, the full interval in order to prepare his argument for indictment against Bella Starr. John Figueroa, the same bearded, silver-tongued DA who had pushed to have Denny formally charged with Augusto's murder, was now attempting to do the same to Bella. The hearing was scheduled for 1:00 PM the next day, and Figueroa had requested the principals involved in the case to assemble for one final briefing before he made his grand entrance in court.

Poor Denny. When he'd finally returned to his home outside of Santa Fe, a tremendous feeling of relief flooded through him. The familiar surroundings, and a return to his daily routine, sapped tension from his neck and shoulders, and he'd slept soundly for almost ten hours that first night. Next morning, over coffee and homemade scones, he'd brought his wife, Pamela, up-to-date on all that had happened to him. He sheepishly apologized for his failure to contact her while he was in jail, expecting a stern rebuke and lecture. Instead, he was treated to a welcome dose of sympathy and understanding. He hadn't appreciated that his gaunt, fatigued appearance conveyed much more about what he'd been through than any number of words could have done.

Within days of his escape from the site of his ordeal, however, Denny was back. Summoned to the Taos County courthouse, he was part of the select group now assembling in a conference room on the second floor of the judicial complex. Denny sighed. He couldn't dispute the necessity of the meeting. The prosecution had to present its case tomorrow and was woefully unprepared. Figueroa had spoken only briefly with Sheriff Wade about the evolution of the charges against Bella. The sheriff had just gone through major surgery to repair his shoulder and, prior to that, he'd lost a lot of blood. He'd been only marginally coherent for days, and, in any event, the sheriff had to admit he didn't know all the facts. Horace Begay had been attending to other business for the past week and had, therefore, been out of touch. It was perhaps ironic that the most celebrated case in the history of Taos County had thus far been ignored by the prosecution. Figueroa was understandably on edge. He was running out of time.

THE MURDER OF Augusto Benavides had suddenly become national news. Hotel rooms were hard to come by as press crews from all the major networks and newspapers descended upon Taos. During the day, reporters camped outside the county jail giving on-site interviews and trying desperately to catch a glimpse of Bella in prison garb. Secondary crews, from the three local stations based in Albuquerque, also hovered around the Stetson ranch on the off chance that Dirk might make a surprise appearance there.

For the past week Mother Nature had been decidedly uncooperative. The volatile fall weather patterns so characteristic of northern New Mexico, had arrived and at the most inopportune time. Strong winds gusting to 50+ miles-per-hour were now in the forecast, and nighttime temperatures had dropped into the low twenties. Undaunted, the huge cadre of intrepid news people assembled in this small southwestern town made the best of their difficult environment. Anchormen wore heavy coats and scarves while conducting their nightly collections of news vignettes from temporary "headquarters" in the public square. Each network had to carefully orient its cameras so as to not reveal the equipment of competitors set up only a few yards away. Minor scuffles over territorial imperative were common, but, come crunch time, each organization conducted its own broadcast in well-oiled, professional fashion.

Overseas, conflict raged in the Middle East where renewed violence of the "Arab Spring" threatened to topple several despotic regimes. The dollar slipped badly in foreign markets as the greedy, bloated government in Washington raised its debt limit yet again, granting itself permission to borrow more paper from the Fed. A mine explosion in northern Kazakhstan buried seventeen men under tons of rock, their air supply dwindling by the minute as frantic rescue efforts began. All around the world events occurred that claimed lives and changed the course of history. From the US news media, however, very little information about these events was presented to the public. The vast majority of airtime, in some cases entire broadcasts, was devoted to one, and only one, story: the indictment of the great American actress, Bella Starr.

During one live update, filmed in front of the county jailhouse, an NBC reporter narrowly escaped injury when a satellite dish, unhinged by a particularly violent wind gust, broke from the roof of her news van and sailed like some giant Frisbee within inches of her carefully coiffed head. The woman didn't notice. She finished her thirty second spiel in the grave, professional manner of a seasoned reporter. However, the camera happened to capture the scene from an angle that made the dish look like a strange, alien spacecraft rushing through the streets of Taos.

"Holy (bleep!), d'jou see that?" an off-camera technician exclaimed as the incident played out. His expletive was barely caught in time by an alert sound man during the fifteen second delay. Aside from a disparaging glance from the reporter, the segment ended without further complications, but the real story had already happened. That small flying disk captured the attention, if not the imagination, of an entire nation. Tweets from thousands of viewers immediately peppered the airwaves with descriptions and speculative explanations of what had transpired. A growing conspiracy felt it was a sign that the impending court appearances and trial transcended mere international interest and had achieved true intergalactic status. NBC at first tried to convince the public that no extraterrestrial contact had occurred, but soon realized the inherent value of the serendipitous event. For the next several days it capitalized on the "flying saucer" incident with slow-motion footage and campy sound bites.

Within hours of the broadcast, a handsome, articulate young man with long, dirty-blonde hair tucked under a foil-lined hat appeared at the door of NBC's local production headquarters. The strange caller ultimately charmed his way into an on-air commentator role as an optimistic and passionate proponent of the likelihood of extraterrestrial contact. In lively debate with experts from NASA and JPL, he presented his case with an upbeat blend of knowledge, skill and humor. Strangely, the only remuneration the fellow sought for his charismatic testimony was a high-definition copy of video footage of the event: something suitable for 'research purposes.'

NBC's ratings soared. Life was grand for the network as preparations continued and tension rose in the days leading up to Bella's first appearance. It was amid this carnival-like atmosphere that the prosecution worked feverishly to finalize its case against Bella.

IT WAS STILL a few minutes before 10:00 and people were filing into the room. Most of the seats around the large central table were occupied. Figueroa held court at the head of the table. A slender, dark-haired woman in a well-tailored suit sat to his immediate left. She had introduced herself simply as Elaine, a recent law school grad who would act as Figueroa's assistant on this case.

Denny and Greg took seats in the back of the room by the exit. Greg was excited to be a part of the proceeding, but Denny'd had too little time to rest and recuperate from his recent ordeal. Having gotten precious little sleep the night before in anticipation of this dreaded trip back to the 'circus' that had become Taos, he was in a hurry to get this over with and get out of town. He looked out the sole window in the room. Clouds covered a melancholy sky that threatened to spit rain. The weather matched his mood. Horace was helping Celia Mondragon with her chair. A bulky flak-jacket showed in outline underneath her sheriff's uniform and the caution with which she sat revealed how much her broken ribs still bothered her. Lieutenant Begay was going the extra mile to make her comfortable. Denny glanced over and raised his eyebrows in an expression of interest. Horace blushed. Yeah, there was romance there all right.

Bobby Wade entered the room looking pale and tired. Surgeons had finished with him only days earlier, and he was supposed to be at home, in bed, recuperating. No one in the room was at all surprised, however, at his attendance. The man was notoriously tough, and his sense of duty demanded he be present when the formal arguments to send Bella to trial were pronounced. Camille Baldonado sat deferentially behind him and to his right. Next to her was an empty chair reserved for Paul Viera.

Paul's attendance had been superseded by a need for manpower to help with crowd control at the building entrance. Minutes earlier, judge Hinojosa had decided to ban cameras from the upcoming hearing. He'd done so in attempt to dampen the chaotic atmosphere that threatened the sanctity of the proceeding. News people were now behaving like sharks at a feeding frenzy, trying to reserve, buy or extort space for their sound crews and sketch artists in the small courtroom.

"The secondary charge of attempted murder is a lock," the DA abruptly began. Introductions were a formality that could be dispensed with; Figueroa was focused on getting started. He did allow himself a brief smile as he looked at Celia and Horace. Tomorrow he would have the victim, a respected police officer, and an eye witness, a respected State policeman, standing beside him in court. For a prosecutor trying a high-profile case, that was almost too much to ask for. "Today's discussion will focus on events that preceded the showdown in Mora. I want to know how you figured out it was Bella who killed Augusto Benavides. I'd also like to remind all of you that tomorrow's hearing isn't a trial; we simply need to present a strong enough argument to ensure that there will be one. Ms. Starr has retained one of the best legal teams in the country and we'll have to be on our toes to counter their efforts. Don't forget, though," he sanctimoniously continued, "Ms. Starr deserves the same consideration as any American citizen accused of a crime. In our system of jurisprudence she's innocent in the eyes of the court until proven guilty."

Denny frowned and picked at a hangnail. "What a crock," he thought. "Despite evidence to the contrary, I was treated as though

I was Hannibal freakin' Lecter, denied bail and shackled everywhere I went." He didn't imagine for one minute that Bella would be granted bail, but her celebrity status would win her creature comforts that were refused to cattle like him. He also didn't like it that the DA was approaching the hearing as though it were some sort of competition. "Shit," Denny thought. "There's nothing I can do but to participate in this game. That's why I'm here."

The criminal justice system most definitely wasn't to Denny's liking. He felt there had to be a better way of handling this stuff. Having watched portions of other celebrated cases such as the O.J. Simpson and Casey Anthony murder trials, he knew how unprofessional and wild things could become. Denny had now experienced the system from two radically different perspectives, and he'd seen enough. In his mind the US didn't have a 'judicial' system, it had a fatally flawed 'legal' system wherein the goal of both sides in a criminal hearing wasn't justice—it was *winning*. Defenders would pull out all the stops to exonerate their clients, while prosecutors did whatever it took to convict them. Theoretically, if each side presents its case well, the facts will all come to light and a responsible judge or jury will adjudicate based upon knowledge, law, reason and precedent. Reality, Denny felt, came up far short of this ideal. Insignificant facts were often elevated in importance if they helped support an argument, while important evidence was often downplayed in an attempt to discard it, if it would harm one's case. Strange scenarios were proposed to account for certain facts, and efforts to obfuscate the truth, accompanied by verbal jousting and theatrics, were promulgated to sway juries. In fact, the seemingly simple process of jury selection was considered to be an art unto itself. When push came to shove, the goal of any 'good' lawyer was to win. Although Denny held Figueroa in contempt because of the way he'd been treated, the man was no different from any other prosecutor. Guilt and innocence were largely irrelevant; it was his job to convict whomever the police arrested.

In Denny's mind it all boiled down to this: is the potential for achieving justice greater under the current adversarial system, or would the chances for corruption and scapegoating be lower if panels of reasonable, impartial experts were elected to both try and judge the accused? Wouldn't a defendant's case be better handled without the grandstanding and wild speculation? He stared out the window again. The majestic peaks of the Sangre de Cristos were almost completely obscured by windblown dust—almost. The mountains beckoned, and pangs of longing shot through him. The joy of being outside doing fieldwork, the smells of fall and the sound of rustling aspens, things he'd experienced barely one week ago, seemed little more than a vague memory from his distant past. The mention of his name interrupted his thoughts.

"It was Mr. Carswell who led us to suspect Bella," Horace Begay was saying in response to something Figueroa had asked. All eyes were now on Denny. Elaine's pen was poised above her legal pad in anticipation of whatever he had to say. He cleared his throat and fidgeted in his seat while his mind switched gears, racing to formulate a coherent explanation of what had transpired.

"It was difficult for me to think clearly right after my arrest," he began. "To have been wrongfully accused of a terrible crime and then denied bail . . . well, let's just say that has an adverse effect on one's cognitive abilities." He glared at both Wade and Figueroa in turn. When neither exhibited any sign of chagrin, he just shook his head and continued. "The land sale had to provide the motivation for the killing. From the strange verbiage in the deed, I knew something was wrong, and I began to suspect that the land that was for sale actually might belong to the Benavides family. There's evidence of an old fence line that runs northwest from the paraje San Cristobal to the Rio Castaneda." He shrugged, "we may never sort out the land ownership issue, but it was obvious that Augusto was causing problems for the killer. Now, in any transaction there are two parties involved: buyer and seller. In this case the buyer, Santa Teresa Pueblo, certainly had strong motivation to consummate the deal. They stood to make millions from casino revenues once they owned the land, and they'd been trying to buy it for several years. The possibility arose that a rogue member of the tribe might have killed Augusto out of exasperation because of his continuing efforts to impede the sale. But Emilio alluded to financial problems that the Benavides's were having. Augusto was too old to manage a profitable number of stock, and Maria's stroke resulted in high medical bills. I became convinced that Augusto was only trying to stop the sale because he knew he was the rightful owner, and I assumed the Pueblo knew that. I further believe he would have been more than willing to sell the land in order to eliminate his financial woes if the deed showed him as owner rather than Dirk. Therefore, it made no sense that anyone associated with the Pueblo would have harbored ill feelings towards Augusto. They just wanted to get the land; it didn't matter from whom they bought it, and they certainly wanted clean title to it. That meant it was most likely someone from the Stetson ranch who was the killer. All of those present at the ranch on the eve of the murder later assembled at the murder scene, and they were questioned at length by Sheriff Wade and his deputy. Of course, I was there too. If anyone else had been present at the ranch when the killing took place, someone would have mentioned it. Therefore, I became convinced that one of the people standing there that evening had killed Augusto less than an hour earlier."

Elaine was taking copious notes while her boss listened calmly. Figueroa's elbows pressed on the table, his lips were pursed against his interlaced fingers in thoughtful contemplation. Denny arose and made his way to a tall coffee urn sitting on a serving cart by the door. He grabbed a large Styrofoam cup and prepped a satisfying blend of sugary java before continuing his narrative. It gave him a modicum of perverse satisfaction to make the DA wait for him to perform some menial task before he revealed the evidence the man would later present in grandiose fashion in the courtroom. He casually stirred his coffee and sat back down.

"When I reflected upon what I'd seen and heard at the murder site, the number of possible suspects boiled down to four. Dirk, and possibly Conrad, were taller than the figure I'd seen assassinate Augusto. It was also difficult to believe that Conrad would have hiked to the summit of Cerro Medio to kill the man. He seemed averse to strenuous activity. If he were plotting to kill

Augusto he'd have found an easier way. And Lily and Emilio?" Denny paused to sip his coffee and thought for a moment. "I have to admit that my dealings with those two indicated they were good people, but gut feelings aren't always reliable. I didn't feel Bella was capable of such malice either until I realized she had lied to Sheriff Wade." He sat up straight and leaned forward, his features hardening as he did so. "She claimed she'd been sunbathing near the pool when the killing occurred. As soon as I recalled her statement, I realized there was a way to either confirm or refute that assertion. Earlier that same morning, before running a traverse to tie remote corners to those I'd already located by GPS, I took a few side shots to the corners of the ranch house and surrounding out-buildings." He looked at Greg before continuing. "A thorough boundary survey doesn't just include the corner monuments that define a parcel. It should also locate notable features on the property such as key landmarks, permanent structures, fences, driveways, known easements, etc. Emilio offered to help run the prism pole while I did that."

When he thought about Emilio, Denny temporarily lost his train of thought. He suddenly felt silly about the way he himself had been treated when the image reappeared of the ranch hand's corpse lying on the dirty barn floor. He looked over at Bobby Wade and lowered his voice. "He, uh, kindly offered to help me even though he had more pressing chores to attend to. Anyway, those measurements allowed Greg and our colleague Bonifacio Romero to use trigonometry to compute the bearing of the east wall of the house. Knowing that the back wall is orthogonal, gave us its bearing as well. The two of them then computed the altitude and azimuth of the Sun at several precise times in the interval between four and five PM on that particular day. Those data, combined with the height of the roof and the patio dimensions, allowed them to determine exactly when the patio went into full shade. As I recall, they found that Bella could not have been sunbathing after 4:34 PM. She maintained that she was out there until after five when Deputy Viera showed up at the front door."

Figueroa drummed his fingers on the table. "So, you basically showed that Bella wasn't sunbathing because the pool area was in shade. And you can prove this? Beyond a shadow of a doubt?" He smiled at his own joke.

Denny nodded patiently. "Horace and I later discussed possible motives she might have had for falsifying the land deed and going so far as to murder Augusto. I remembered seeing Bella talking to a man, a guest from the set of Dirk's movie, on several occasions. They seemed to be rather close." He looked at lieutenant Begay as a cue for him to take over. Horace, who looked about as pleased as Denny to be cooped up in a meeting, gave an almost imperceptible nod and picked up where Denny left off.

"Both Sheriff Wade and I made trips to the movie set down in Abiquiu to talk to the fellow." He smiled, "Denny told us to look for a bald-headed man named Andre. He wasn't hard to find. Anyway, he was very candid about certain aspects of Bella's life. He confirmed suspicions Denny harbored regarding her financial status, and, most importantly, he told us where he'd seen her using her cell phone."

"That would be the one used to call Augusto on the day of the murder—when she set him up?" It was Figueroa's assistant, Elaine, who spoke. She was busy flipping through the pages of a yellow legal pad and hadn't even looked up.

Horace nodded. The woman was silent as she scrawled a few notes in the margin of a page. When she finished, Denny cleared his throat. "Little more than a week ago, I lay in a cell just down the street from here, trying to think of a legitimate and compelling reason why Bella Starr would be driven to murder. The fact that her alibi was a lie was, at the time, incongruous. She had no obvious motive. But there were signs, nonetheless, of a dark side to her seemingly perfect life."

Sheriff Wade and Horace Begay were the ones who had delved into Bella's affairs, but glances of invitation from Denny to continue the narrative were met with expressionless stares. He knew they wanted him to be the one to brief the legal team in the details of what they'd found. If it weren't for Denny's persuasive arguments, the facts would never have come to light. There was an unstated understanding between the two law enforcement officers: Denny Carswell was the one who'd solved this crime. Sheriff Wade in particular had come to fully appreciate Denny's efforts during his recuperative stay in the hospital. Like his newfound friend in the State Police Department, Wade had become an admirer of the old surveyor and his uncanny knack for uncovering the truth. Denny had earned the right to let the world know how he'd unraveled the facts and brought Bella to justice.

"Although she was outgoing, friendly and loquacious, Bella was also manipulative and terribly insecure. She desperately sought recognition and approval from all the men in her life—approval and support that weren't always forthcoming from those who mattered most: her husband and influential figures in the film industry." He paused to sip coffee and stared blankly at the far wall, deep in thought about the fascinating woman who had been driven to such extreme ends. "Bella was unable to cope with her senescent beauty; it compelled her to undergo multiple cosmetic surgeries in vain, valiant efforts to subvert the aging process. I began to suspect that expensive medical procedures, lack of employment, and the need to maintain an extravagant lifestyle had driven her into poverty and made her jaded and hateful of a society that worshiped youth and beauty." He broke from his reverie and turned his attention to Elaine, the lawyer who was recording what he said. "Dirk and Bella were married in 2004, nine years after Dirk bought the ranch." He shrugged, "As I suspected, Dirk and Bella had a prenuptial agreement. Dirk's individual assets, including the ranch, were protected should they divorce. Bella was broke and desperate; divorce would have put her in the poorhouse, and it could have happened at any time. Hell, I saw Dirk openly drooling over his latest fling when I met him at that party. Thinking back to the look Bella gave him that night, I'm surprised he didn't meet the same fate as Augusto.

"Several years ago, a combination of cunning intuition and local events enabled Bella to hatch an ingenious plot to gain a financial windfall at the expense of Augusto and Maria Benavides.

When Santa Teresa Pueblo expressed interest in the bottom land in the northwestern part of the Benavides property, she looked at Dirk's deed and came up with a clever way of falsifying the record to make it look like Dirk, not the Benavides family, owned the land coveted by the tribe. It was simple, and I must admit, brilliant. The last course in the metes and bounds description in the survey attached to Dirk's deed runs from the paraje San Cristobal, marked by a monument at the top of Cerro Medio, to the Rio Castaneda. The direction and length of that line, however, are given on the page she tore from the deed. That page also contained Antonio Roybal's plat of the property. The plat, of course, would also have shown the course of that line, clearly indicating that the land belonged to Benavides. Without that crucial page, it's unclear exactly where the southwestern corner of Dirk's ranch is located."

Denny was becoming animated as he speculated on how the survey and deed had been altered. This was his world. Only an experienced surveyor could have dissected the scheme dreamed up by Bella Starr, a plan that led to murder when it began to unravel at the last minute. The legal and law enforcement representatives in the room would never have solved this crime without him, Boniface, and Greg, and Denny began to wonder if some sort of divine intervention was responsible for his incarceration and his resulting desperate need to find the truth. Had all of this happened in response to Maria's prayers?

"The bearing of the old fence that I saw from the top of Cerro Medio, runs northwest towards the town of Arenas Rojas for a distance of about 3000 feet, ending at the Rio Castaneda. But the Rio curves south and west from the town, paralleling the highway as it winds around, defining the western border of Augusto and Maria's spread. It so happens that the radius of the arc of the arroyo, as measured from the paraje, is reasonably constant at 3000 feet or so. Bella simply moved the re-bar corner to another point in the stream further to the southwest. The result is that a portion of land shaped like a slice of pie, land the pueblo sought to purchase, would appear to belong to Dirk Stetson. Bella's plan probably hatched in 2009 when she learned of the pueblo's intentions. My suspicion is she learned about it from Emilio, who was close friends with several tribal leaders. It was Maria's stroke, however, that undoubtedly spurred Bella into action. She had ample time and opportunity to move the corner stake, remove the page from the deed, and, later, to pay for another survey while Augusto was caring for his wife. Augusto probably didn't know what was going on until he noticed the new fence being erected—the one along the false property line 'verified' by Josh Marquez's survey. The pueblo was talking to Dirk's representatives of course, because, according to the recently falsified public documents, he was the owner of record. The court challenge brought by Augusto was rejected based upon Josh's expert testimony which, of necessity, centered on the forged deed."

Greg Martinez spoke up, "Okay, in general I get what went on, but I have a question. Wouldn't Augusto have had a copy of the Stetson deed, or at least a copy of Antonio Roybal's survey? After all, he's the one who sold the property."

"I'm sure he did—up to a point. But by the time he learned of the altered deed in the clerk's office, his copy would have been either altered or stolen. You see, Greg, Maria and Augusto were simple, trusting people. They had given Dirk and Bella a key to their home—to keep an eye on the place in the event they went on vacation or in case of emergency. Lots of people have arrangements like that with their neighbors. Bella therefore had ready access to Augusto's records, and I'll bet it never dawned on him that Bella was behind all this. If he had had any reason to suspect her of wrongdoing, he wouldn't have agreed to meet her at the paraje the evening she shot him."

"Ah, that brings up the issue of the telephone," Figueroa interjected. "How exactly did you recover it?"

Denny was tired of sitting. He got up and walked over to the window, near to where Elaine was sitting. Alternating his gaze between her notes and activity taking place outside, he continued. "Bella's arrest was the result of a carefully choreographed series of events beginning with issuance of a warrant by Judge Hinojosa. As I understand it, the Judge readily accepted the evidence that Bella had lied to Sheriff Wade, and he was also quite familiar with the controversy surrounding the land deal. He was the one who ruled in favor of Dirk in '09 after the Stetson's put up that fence. Augusto hadn't been happy with that decision, of course, and he was recently in the process of resurrecting the issue. Hinojosa had to know it was just too much of a coincidence that Augusto was murdered right before the land deal was to be consummated. Anyway, it's easy to see why he came through with the search warrant and why he allowed us to carry out the sting operation in Mora.

"What transpired thereafter is pretty straightforward. Horace got the number of the only caller to the Benavides house on the day of the murder. Because it came from one of those pay-as-you-go units, it was impossible to trace, but we knew it probably belonged to the killer. Once Sheriff Wade had the warrant in hand, a search could be conducted, but the Stetson ranch house is a big place; it would have taken many man-hours to pick through it successfully."

Denny was on a roll again, deep in thought as he concisely related the exquisitely coordinated plan that had led to Bella's arrest. Everyone in the room sat spellbound while he talked, even those who had participated in what had taken place.

"Like I said, on several occasions as I walked around the Stetson ranch conducting my survey, I noticed Bella talking to a bald-headed man. The two seemed to be very close, so I felt he would be a good person for the police to interview."

Bobby Wade winced as he repositioned himself in his chair. Holding the sling containing his injured arm with his good hand, he nodded at Denny. "Yeah, 'ole Denny's hunch was right. That fellow, Bivens, he's the director of Dirk's new film, 'Conestoga Highway.' I drove down to Abiquiu that day to talk to him. Didn't think much of me interrupting his job, but I's able to persuade him that it'd be worth his time to answer a few questions. *Heh*, I didn't even need to get tough with him. One of the steers they was usin' on the set got spooked and took off. Nobody there knew what to do; they all just stood around with their heads up their corn bins, so I jumped on a horse, rode out and escorted the animal back to its pen. That loosened up Andre's tongue. He told me if I'd done that three months ago, he'd a asked me to be the star of the film. Seems he don't think much of 'ole Dirk." Wade paused for a moment and slowly shook his head. "Now, I s'pose I shouldn't judge people like I do, but dang it, Bella's a married woman, and Bivens told me the two of them were 'bumpin' uglies' right there in her house, in one of the guest rooms!" He suddenly turned sheepish and looked in turn at Elaine, the note taker, and Sheriff Mondragon, the two women in the room. With a nod he said, "My apologies, ladies," before addressing Figueroa once again. "Well, Bivens told me that after they'd uh, finished their, uh, business, right before he left to drive back to the set, she left the room for a while and then returned holding a cell phone. When they walked from there to the parking lot she wasn't carryin' nothin'. To make a long story short, I called up Camille and Paul, told 'em where to search. They found the phone stashed up on a shelf in a closet of that room. The phone number matched the one that called Augusto that day, and the time Bivens estimated he left the ranch coincides, more-or-less, with the time of the call."

Figueroa nodded in thought. "And you conducted the search while Bella was in Mora, correct?"

"When she was en route," Denny said, his tone becoming bored. He was getting restless. Figueroa now had all he needed, and more.

The night before, Boniface had called Denny and excitedly told him he had important news. When Denny mentioned he and Greg had an appointment in Taos, they had agreed to meet in Arenas Rojas later in the day. Boniface was willing to make the two hour drive from his home in Pecos because he also wanted to talk to Maria Benavides. They'd arranged to meet at Alice's Restaurant at 4:00 PM. It was now almost 2:00. With a sigh, Denny hurriedly answered the DA's question hoping he'd be able to leave soon.

"Sheriff Wade thought it best to search the Stetson residence without Bella around, so his deputies waited for Greg Martinez to call when he saw Bella leave for the Mora County courthouse. Greg was absolutely central to the operation in Mora. Hell, he's the one who discovered that the deed was missing a page."

Denny looked at Greg and raised an eyebrow. Greg had been Denny's apprentice for three years now and understood the gesture. It was his job to elaborate on what had happened at the Stetson ranch when he went there under the pretense of finishing the survey.

"The boss here came up with the idea of forcing the killer's hand. He had me draw up the fake deed containing the missing page. I was surreptitiously to spread word of its 'existence' at the Mora courthouse. Before going to the Stetson place, I gave the deed to lieutenant Begay. He took it to Mora and set up everything there."

Greg's eyes drifted to the pretty young policewoman seated beside Horace. He knew all about what had happened in Mora, about how she'd been shot. A pang of guilt came over him. He felt partially responsible for her injury just as he did for what had happened to poor Emilio. He was learning at a young age, perhaps too young, how fragile life can be.

"As far as my job at the ranch was concerned, Denny told me to announce I'd be going to Mora the following morning to examine the deed. He made me promise to make certain Bella heard me and to make sure I spoke about it with other people present."

Denny interrupted, "That was Horace's idea. He felt Bella might have been a threat to Greg if she thought he and I were the only ones immediately aware of the existence of another copy of the deed. You see, as a surveyor, Greg was capable of creating problems for the land sale. The only others who posed a similar threat were Augusto and me. But after killing Augusto and framing me for the murder, we'd both been effectively neutered. Killing Greg would have put a stop to any foreseeable analysis of survey data before the sale went through. Mention of the deed in front of witnesses was brilliant because it made further killing too risky. The only alternative left to Bella was to get to the deed in Mora—remove the last remaining evidence—before Greg did."

"So, Greg planted the information and watched while Bella took the bait and left for Mora. He then phoned the Sheriff's office so that they could serve the warrant and search the Stetson premises. Is that about it?" Figueroa asked.

"Yep," Denny replied making a demonstrative glance at his watch. "I believe you know the rest."

The DA clasped his hands behind his head and closed his eyes in thought. This case was evolving into a prosecutor's dream. He was about to present a case against one of the most well-known celebrities in film, in front of the national media, and his arguments were absolutely air-tight. In the legal world, John Figueroa had just won the freakin' lottery. Abruptly he sat up straight, eyes bright, and said in energized voice, "That's good, really good. I want to thank all of you for your help in this matter. We have a strong case. The phone, the altered deed, the attempt on Sheriff Mondragon's life, Bella's financial problems, the ballistics data—all of these facts in aggregate should be enough to force either a plea bargain or a trial."

The meeting adjourned and everyone prepared to leave. As he and Greg arose, however, Denny did an abrupt double-take. "Ballistics data? Why would the fact that my gun, the one used to kill Augusto, constitute evidence against Bella?"

Figueroa smiled, "Your gun may have been the murder weapon, Mr. Carswell, but the bullets used to kill Augusto didn't match the shell casings found in your truck. They came from a box of cartridges found in the barn at the Stetson ranch. Emilio noticed it had been opened and gave it to Sheriff Wade. He thought it might help to exonerate you."

Denny had nearly forgotten that he had had no ammunition in his vehicle. Even the gun itself shouldn't have been there. He'd simply forgotten he'd stashed it behind his seat after target shooting with friends. Actually, the thought had crossed his mind about the ammo, but he hadn't attached any significance to the fact. 'Twenty-two caliber long rifle' was the most prevalent rifle round in the country, but he suddenly realized that, while the cartridge was standard, bullet types varied widely. He mentally kicked himself; he should have known! "So, you found spent shell casings in my truck? That's quite understandable."

"The bullets that killed Augusto were jacketed hollow points," Figueroa continued. "The spent cartridges in your truck were non-jacketed target rounds. A database search indicated they didn't even come from the same manufacturer."

Denny paused to contemplate this new information. He was chagrinned that the subject of ammunition had escaped him, but then another point came to mind, one that stirred anger within him. "You had to have known this since before my arraignment," Denny challenged.

"Why do you say that?"

"Let's not bullshit one another, shall we?" he said vehemently." The room fell silent as Denny vented. "As I recall, when I was arraigned you read lots of gratuitous detail from the autopsy report, and then pressed on about how ballistics showed that the bullets were fired from my gun. You had to have known about the ammunition issue back then but didn't mention it because it might have jeopardized your argument to deny bail. Correct? And when were you going to reveal this to me or to my lawyer?"

"When the time was right you would have been informed. I was just doing my job . . . ."

"Just doing your job, huh? Is withholding evidence that may damage your case part of that job?"

"Now wait just a minute . . . ."

Denny cut him off. His jaw was clenched and his hands were balled into fists, "Someday I hope some smartass lawyer tries to railroad you into a murder conviction. Shit, arguing with you is a waste of my time. I've got to get some air." He spat the words as he grabbed his briefcase and stalked out into the hallway.

Right away Denny regretted losing his cool. Snapping at Figueroa wasn't going to do anyone any good. The guy wasn't that bad. He was about to turn around and head back to apologize, but something held him back. There was a modicum of truth in what he'd said. Maybe people in the criminal justice system needed a kick in the pants now and then—at least whenever they falsely accused innocent people.

A shadow cast by someone nearby blocked the overhead light. Denny turned and saw Horace Begay standing a step behind him. It unnerved Denny whenever Horace did that. The big man moved like a cat, never making a sound; he was always unintentionally sneaking up on people. Denny let out a deep breath and his shoulders slumped, "Sorry about that, Horace. I don't know why I lost it. It's not like me."

"I know." A pause ensued. "It didn't surprise me though, Denny. I've seen it before when officers or victims come through a horrific experience like being shot at or watching a loved one die. I'm no psychiatrist, but it's pretty obvious you're still under a lot of stress. I believe it's what they call PTSD, 'post-traumatic stress disorder.' A lot of combat vets suffer from it, sometimes for years after going through a terrifying event. Spending a week in jail, facing a possible death sentence, and watching circumstantial evidence mount against you—that's enough to make anyone buckle emotionally."

Denny allowed himself a smile of relief. "Thanks, Horace. Really. It helps a lot to know that someone sympathizes with what I went through." He changed the subject. "So, where have you been for the past week or so? Figueroa was darn near pissin' his pants in there trying to learn what happened with Bella. I figured you'd have provided him most of the information before today."

"Contrary to what the press would have people believe, there are other important cases going on. I've been busy in Albuquerque and Santa Fe." He hesitated for emphasis, "Three nights ago we caught the guy who murdered the boy from Taos Pueblo."

Denny perked up at the comment and broke into a broad smile, "Horace, that's great news!—really important news. Why haven't I heard about it?"

Horace chuckled, "You know as well as I do that there's only one story the press gives a shit about. But Bobby and the Pueblo sure are happy." His tone grew more serious, "I'm really pleased as well. Murder isn't a particularly common crime, but I've investigated my share. Every one of them is sad, unpleasant business, but when the victim is a child . . . ." After a contemplative silence he said, "Hey, I've got another question for my favorite crime-solving surveyor: was Conrad collaborating with Bella on that land deal? If not, well, I'll be darned if I can think of any reason he'd have to kill Emilio."

Greg Martinez showed up and pointed to his watch. They had to rendezvous with Boniface and had just enough time to get to Arenas Rojas by 4:00. "The short answer, Horace, is: I don't believe Conrad and Bella were in cahoots. At the moment the motive for his rampage is unknown, but Boniface suspects something. He feels that Conrad's aunt, Maria Benavides, can shed light on things. We're going to speak with her this evening. I'll keep you posted." He and Greg strode down the hallway, but at the entrance to the stairwell, Denny did a double take. "Hey, Horace," he yelled back, "whatever happened to Conrad? He's not being held in the Taos jail; is he?"

Begay and Celia Mondragon were now making their way slowly towards the stairs. Horace wore a huge grin, in part due to Celia's presence and also because of Denny's question. "First thing Judge Hinojosa did following Conrad's arrest was to request that I transport him to the State Penitentiary outside Santa Fe."

"That's understandable, I suppose. With the press hanging from the rooftops, it probably pays not to have other prisoners around."

"Oh, that's not the reason," Horace laughed. "Hinojosa's no dummy. He knows Bobby Wade well enough. Just consider that Sussman shot Wade right after killing Wade's best friend and you can understand the judge's concern. Chances are 'ole Conrad might suffer some sort of 'accident' were he to stay here, like maybe getting shot while trying to escape or undergoing sudden defenestration from a third floor holding cell. The temptation for vigilante justice might be a bit much for the Sheriff to resist. Besides, Conrad, like Bella, is considered to be innocent until proven guilty. We wouldn't want to deny him his day in court."

XVII

High winds generated by 'virga,' rain that evaporates before reaching the ground, buffeted Denny's truck as he and Greg made the drive to Arenas Rojas. The sign outside Alice's Restaurant rocked back and forth in the stronger gusts, and when Denny opened the spring-tensioned entry door, it blew out of his hand and banged against the outer wall. Despite the inhospitable weather, the restaurant was packed—quite unusual for a weeknight. Jenny spotted him immediately and waved, her indefatigable ebullience unfettered by her high work load.

"My, you're busy," Denny mentioned as she approached, coffee pot in one hand and dirty dishes in the other, "what gives?"

"Ever since they arrested Bella we've been swamped; a lot of TV people are spilling over from Taos. Local customers are a bit put out by all the ruckus, but I can't complain. These out-of-town folks are pretty good tippers. They must be on expense accounts or something. Hey, your friends are in that booth in the far corner." With her hands full, Jenny motioned with a nod towards the back of the diner. Denny looked up and spotted a beckoning wave from Bonifacio Romero. It was a pleasant surprise to see that Rosa had accompanied him.

"Hola, niño. Hola, joven," Boniface said to his friends as they took their seats.

Greg laughed, "Do I really look that young?"

"My boy, when you deal with senile elders like me, you got to go with the flow. You can call me 'El Pedo Viejo' (The Old Fart) if it suits you."

"Now, Boniface," Rosa admonished.

"What?" he exclaimed in mock protest, "I'm not cussin', just speaking the ugly truth."

"Don't forget," his wife continued, "I'm as old as you are. Don't you think it's insulting . . . ."

"You are? But you look so young! After all these years I would never have guessed . . . ." He put his arm around her and gave her a peck on the cheek. Denny grinned. He was used to this banter from the Romeros and reveled in the fact that, even though they were pushing eighty, they loved life and one another to their fullest.

"I think niño here could use some pie, and I need coffee," Denny said.

"Comin' right up." Boniface caught Jenny's attention and she floated over to their table.

"What'll it be, Boniface?"

"Coffee there, a large slice of your fine peach pie over there, and refills for me and my beautiful bride."

"Sure thing," she replied with a larger than normal smile.

When Jenny was gone, Denny looked at his mentor, "'Boniface'? Since when have you and Jenny been on a first name basis?"

"We've been here several times, most recently right after Augusto's funeral service."

All feeling of levity left the table at the mention of the funeral. "How's Maria holding up?" Denny asked somberly.

"She's doing well. She's expecting us in about an hour, after dinner."

Jenny arrived with Denny's coffee. She'd brought more tea for Rosa and refilled Boniface's mug with a wink and a smile. "I'll be right back with your pie, good lookin'," she said to Greg, making him blush.

"Well, what's this news you promised us," Denny asked, tearing open two sugar packets and adding the contents to his brew.

"Antonio Roybal has a sister."

"And?"

Boniface wore a sly grin, "Had a heck of a time finding her."

"Let me guess," Greg said, "She's married and has a different last name."

"Bingo. That's partly why Josh Marquez couldn't follow up on what he'd done—that and the fact Antonio lived way down the valley." Boniface slowly sipped his coffee, luxuriating in Denny's suspense. "Now, Greg, one of the advantages of getting old—well, maybe the only advantage—is that you get to know lots of people. Rosa and I paid visits to a number of old acquaintances, and we found her. Name is now Naranjo. She's widowed and lives in Alcalde." He hesitated.

"And?" Denny looked ready to explode out of his chair.

Boniface's smile widened as he lifted his jacket from the seat beside him and pulled something out of a pocket. It was an old yellow field book, of the standard size carried by nearly all surveyors. The edges were dog-eared, the lower half stained and with wrinkled pages. Embossed with the manufacturer's initials, K&E, the front was labeled in waterproof marker, '95-III. Post-It notes had recently been added to mark selected pages. Boniface deposited it with great ceremony in front of Denny.

Denny accepted the book with deference and looked up at Boniface. "He had a brother too," Boniface continued, "but Josephina, his sister, kept most of Antonio's stuff after he passed away. That first page I marked—you'll find that particularly interesting."

The entry bore the date August fourteenth. The left hand page contained angles and distances recorded in neat, small script carefully arranged in columns. On the facing page was a carefully drawn portion of the western half of Dirk Stetson's property. From a point labeled 'paraje S-C' began a line that ran up the page to a terminus in the center of a sketch of the Rio Castaneda. Above the boundary was recorded its bearing, 'N 5-34 W', and below it, a distance, '2984.50 ft'. Denny looked first at Bonifacio and then pointed out the feature to Greg. "So, what do you think?' he asked.

Greg studied the drawing. "Looks like the true western boundary of Dirk's parcel runs along that old fence line you spotted." He shrugged, "Augusto and Maria Benavides are the rightful owners of the land to the west, the spread that's for sale. You were right, Boss."

"I don't know how to thank you, Boniface" Said Denny. Your ability to research survey data like this is amazing."

"Nonsense," came the modest reply, "I'm so old, and the communities in this part of the State are so small, it's almost impossible ***not*** to find a friend of a friend who has what I'm looking for. Another thing that helped was that Rosa and Josephina are distantly related, cousins or something."

"To her husband," Rosa said, "My grandfather's sister, my great aunt, married a Naranjo. And," she smiled, "Josephina and Antonio have a younger brother, Enrique, who's moved away. I dated an Enrique Naranjo in high school. Now, Naranjo is a common surname around here, but it might have been him." She took a sip of tea, "He was cute."

"Well, then, it's a good thing I came along," Boniface remarked. "Just think, if you'd stayed with Enrique, you wouldn't be married to a big shot successful surveyor like me."

Rosa put down her cup. Her eyes twinkled. "Boniface, let me tell you something. I heard a story once about President Truman. It seems he and his wife were driving through Independence, Missouri one fine summer day when they passed a run-down gas station where an old gentleman was bent over servicing a car. Bess turned to her husband and said, 'See that man working over there, I dated him in school.' Harry replied, 'Just think, Bess, if you'd married him, you wouldn't have become First Lady.' To which Bess replied, 'Oh yes, I would have'."

The men at the table roared in laughter, none more so than Boniface. Amused looks came their way from the other patrons of the establishment. Rosa remained impossibly deadpan, continuing to sip her tea, exhibiting little more than a sly smile. Once Denny regained his composure, he managed to stammer, "What I love most about you, Rosa, is that you're the one person in the world who can put this guy in his place."

MARIA WELCOMED HER four visitors warmly. She and Rosa shared a long hug before everyone took his seat in the cramped room. Boniface and Rosa sat at the end of the bed, Denny pulled up a chair from over by the window, and Greg hopped up onto a low dresser by the door. Maria slowly made her way to her favorite rocking chair—strategically placed in front of her devotional statue of 'Our lady of Guadalupe.' A votive candle flickered in the gentle breeze emanating from a ceiling vent. Maria and Denny had never met, but each felt as though he/she knew the other. Rosa formally introduced them and Denny expressed sincere condolences for what had happened to Augusto.

Greg remarked, "You're looking good, Señora Benavides, getting around without your cane. I see marked improvement since the last time I was here."

Maria beamed, "Thank you, Greg. Yes, I am much stronger. Soon they will let me go home, I think."

Rosa and Boniface shared a knowing look. Both knew how much Maria wanted to go back to her beloved ranch, but despite her recovery from the stroke, she wouldn't be able to live there alone. She was going to need help.

Rosa broached the subject of Augusto's funeral and burial, a topic that, surprisingly, she knew would lift Maria's spirits. The woman was very devout. It had been a simple but reverent ceremony, and her husband had been laid to rest in a small family plot at the ranch.

"God has blessed my Augusto with a beautiful resting place, has he not?" Maria said contentedly. "I am old and I know we will be together soon, side by side, overlooking our valley." She sighed and closed her eyes. She seemed at peace, wistfully musing about the future. When the feeling seemed to pass, Boniface cleared his throat.

"Maria, there is another purpose to our visit. First of all, I have good news about the ranch." He summarized the survey findings, promising to restore her title to the land Bella had tried to steal. "It was admirable the way Augusto never gave up trying to protect what was rightfully yours."

"Yes," Maria said proudly, "My Augusto was a good man. I think even if he'd known it would take his life, he still would have fought for what he believed."

"There is something else as well," Boniface said. He leaned forward and spoke softly, "You mentioned to Rosa and me, quite some time ago, that you had papers showing title to your land. Furthermore, you said those papers belonged to your great grandfather. Maria, your family bought the ranch from the government long after your great grandfather had died."

Greg weighed in, "Yes, you mentioned that there were copies at your house when I, Emilio and Conrad were here last week, right before . . . ." Suddenly, Rosa broke into a frightful coughing fit, interrupting Greg and causing alarm among the others. As Boniface patted her back, she recovered immediately and shot him a warning glance. Maria didn't yet know of her nephew's terrible crime—the way he'd brutally murdered her kind and generous neighbor.

Boniface took the hint. "Oh, I remember you telling me about that, Greg. So, Maria, do you have any idea where Augusto put those papers, in a safe deposit box perhaps?"

Maria took on a worried look. She began wringing her hands as though she'd been asked to do something illegal. Turning her pale countenance towards 'Our Lady of Guadalupe,' she made the sign of the cross and then murmured a short prayer. Finally, following a long, palpitating silence, she nodded and said to no one in particular, "Yes, it is time." Arising from her chair with difficulty, but also with grim determination, she walked to her devotional statue and, making the sign of the cross once again, blew out the candle and attempted to turn the heavy icon. Greg leaped down from his perch atop the dresser to help. "Gracias, Greg," she said as her gnarled hands fumbled with a small clasp at the back of the pedestal. Opening a small door, she reached in and pulled out what looked like a scroll. Greg carried the document to Boniface and then helped Maria return to her seat.

The item was wrapped with a worn, deep-purple ribbon which Boniface untied before carefully unrolling what turned out to be a sheet of parchment. Several old pieces of paper were rolled up inside the vellum, which Boniface gently set aside; the outer document was what commanded his attention. All eyes were on the old surveyor as he studied the document. Color gradually drained from his face, and soon his hands began to tremble. He finally looked up at Maria and managed to say in a hoarse voice, "Dios Mio!"

THE CALL WASN'T ENTIRELY unexpected. Santa Teresa Pueblo was getting antsy once again. Denny couldn't blame them; in fact, he felt they were being quite reasonable. That was something he admired about the attitudes of many Native Americans he'd met over the years—life was better if one didn't push the pace too much. After all that had happened at Dirk's ranch, the Pueblo had been kind to those involved and had backed off. But the call he'd received last night from Mr. Hawkfeather had had an understandable hint of exasperation to it.

Denny rinsed his coffee mug in the sink and stretched. He popped the last of Pam's marvelous blueberry muffin into his mouth, brushed residual crumbs from the countertop into one hand and put them into the trash. One of his wife's pet peeves was a dirty kitchen. She wasn't so much a clean freak as she was paranoid about mice. Grabbing his lunch, his thermos and his worn, faded backpack, he walked out to his truck in the early morning darkness. It was now mid-November and the nights had been clear and cold. His hands froze as he warmed up the vehicle and scraped hoarfrost from the windshield and side windows. Despite the cold, the first snow of the season had yet to fall, and Denny wanted to finish this job before a layer of the white stuff made his work more difficult.

Boniface would meet him at the Stetson ranch at 8 AM. As he backed out of his driveway and began the pre-dawn drive towards the highway, Denny wondered what his old mentor had been up to. Boniface hadn't been home for most of the past week. Rosa had answered Denny's calls to their house and told him that her husband was spending all of his time traveling back and forth between Taos and Albuquerque. Boniface himself had finally answered the phone last night, and when Denny mentioned his need to finish the Stetson/Benavides survey, the old man asked to go along. "Oh well," Denny mused, "as always, when Boniface is ready, he'll tell me what's been going on."

THE TWO MET at the front gate of the ranch at five minutes before eight. "Where's Greg?" Boniface asked. He was surprised that Denny's apprentice hadn't accompanied his boss.

"He's stuck in Santa Fe," Denny replied. "He, Lily, and Xeno are giving testimony at Conrad's preliminary hearing. When we last spoke, he told me they should finish this morning and that, if things go well, he'll join us sometime around noon. For what it's worth, Greg says Conrad's lawyers are really pushing for a plea deal. However, it seems Conrad himself still thinks he can win acquittal at the preliminary hearing despite the evidence against him, and if he doesn't, he wants his case to go to trial."

Boniface scowled, "Sounds like that fucker's as dumb as he is guilty—no wonder Augusto didn't trust him."

"Why do you say that?"

"I'll explain later when you buy me lunch. There's more to this story than meets the eye. Until then, are we just gonna stand around or are we gonna do some surveyin'?"

Denny knew his curiosity would have to wait. With a sigh he opened the gate and led the way to Dirk's house.

Clouds were beginning to form to the north and east. Boniface parked in Dirk's driveway and they took Denny's truck up the rugged two-track road towards Cerro Medio. At the base of the hill, Denny parked mid-way between the paraje San Cristobal monument and the nearest corner to the east. The weather report for the Taos area called for increasing clouds, wind, and a chance of afternoon flurries. The early start should get them off the hill and out of the wind in short order. As Denny rummaged in his truck for his GPS supplies, a cold gust pierced his fleece jacket sending a chill through his entire body. He extracted a windbreaker from behind the seat and donned it under the watchful eye of his colleague. "What?" Denny said, noting the dour expression on Boniface's face. "Just 'cause you're immune to hurricane winds and arctic cold doesn't mean I have to suffer."

"Oh, I don't mind that you're a wimp, but what's with that satellite receiver?"

"I've calculated the coordinates of the displaced Roybal corner. While I take the rover pole down to the arroyo, you . . . ."

"That's not real surveyin'," Boniface said with a frown, "It's called standing around with a pole up your ass."

Denny begged to differ but refrained from comment. He knew better than to argue with the old curmudgeon: sentiments and habits die hard. With a sigh of resignation he replaced the suitcases that contained his GPS equipment and began to ready his total station for the short hike to the paraje. As he handed a prism pole to Boniface the icy stare continued. "What now?"

"Fancy laser EDMs, optical encoders . . . hell, we don't need 'em for this job—takes all the skill out of measurement; takes all the fun out of the work."

Denny suppressed a smile, "That may be true, Pedo Viejo, but I personally don't feel like running chain for almost 3000 feet in this wind. What do you say we take the easy way out this time?" With a disapproving shake of the head, his partner began to walk briskly across the ridge towards the point where he'd give Denny a back sight. These sorts of 'arguments' were routine. Boniface used and understood the modern tools of the trade; he just liked to complain at every opportunity.

The old surveyors enjoyed one another's company as they hiked the ranch property taking readings. At the site where Antonio Roybal had set the northwest corner of Dirk's property, the one Bella moved, they discovered a wealth of evidence of corruption. An old, broken piece of lath, about two feet long, lay partially covered with dirt several feet away. Amidst a pile of small rocks were shards of discolored, embrittled, flagging, and, within a few tenths of a foot of where Roybal's notes referenced the corner, there remained a depression, an indication that a small hole in the ground had once been present.

"Well, that's it," Denny said once they'd put their gear away. "I'll prepare a plat to append to Dirk's deed and get Judge Hinojosa to sign it. With the physical evidence we've found and Roybal's original field notes, that should settle the ownership issue once and for all. And those points we set on Maria's behalf define precisely what she wants to sell to the pueblo."

In his travels of the previous week, Boniface had acted as Maria's representative in negotiations with the neighboring tribe. He'd also been kind enough to counsel her about her ranch finances, pointing out that without income from the ranch sale she wouldn't be able to pay taxes on the remaining holdings for much longer. The Romeros had formed a close bond with the unfortunate widow. Realizing that she needed help sorting out her affairs, they had spent considerable time working to make her one remaining dream a reality: that of moving back onto her beloved ranch.

Denny decided to pay a final courtesy visit to Dirk. The star of 'Conestoga Highway' was on break for a few days while the rest of the film crew worked feverishly to prepare their final set. Denny needed to tell him that an amended plat of his parcel would soon be available, and he didn't want the actor to be alarmed when construction equipment began clearing ground for the casino along the western boundary of his ranch.

It was Larry who answered the door when Denny knocked. The ranch manager exhibited markedly less enthusiasm than he'd had at their first meeting. "Hello, Larry. I was wondering if I might have a word with your boss. We've finished with the survey."

"That's great, Mr. Carswell," Larry replied with a touch of condescension, "I'll let Mr. Stetson know, not that it really matters."

"I'd really like to tell him in person . . . ."

"He's unavailable right now. I said I'd tell him for you and I will. I'm the one you should be dealing with anyway. Your check will be in the mail within the week." Larry abruptly closed the door, leaving Denny standing uncomfortably on the stoop.

"What an arrogant bastard," Denny murmured. A flash of anger came over him. After what he'd endured on this job, the man could at least be a bit more gracious. Another thought then sprang to mind, one born of a conversation he'd once had with Emilio. Poised to knock on the door again, Denny stopped himself. 'No, there's a better way to handle this,' he thought with a smirk. 'We'll be seeing one another soon, Mr. Cooper.'

Greg Martinez was pulling into the parking lot as Denny prepared to leave. Behind him was a patrol car. Sheriff Wade had driven Lily and Xeno to Santa Fe for Conrad's hearing. The entourage had just returned.

"I missed out on the survey," Greg noted with a wan smile as his boss approached.

"Wasn't much to it," Denny shrugged. "How did the hearing go? Is your friend Conrad a free man?" Denny's tone was light-hearted, but the response from his normally cheerful assistant was uncharacteristically serious.

"Not a chance. The SOB argued his own defense while four other lawyers sat at his side nodding assertively. The DA thought it ironic that none of Conrad's support crew were employees of his dad's firm, 'Sussman and Poole.' The theatrics were impressive. At one point Conrad tried to pass off Emilio's killing as an unfortunate act of self-defense. 'Ole Sheriff Wade about blew a fuse at that. It's a good thing he couldn't take his gun into the hearing room or Conrad would be pushing up daisies, without a doubt."

"It sounds as though your morning was anything but boring."

"Yeah, that's a good way of understating it. We nailed his ass on the kidnapping and attempted murder charges, though. Lily, Sheriff Wade, Xeno and I all testified to what he'd done. We tore him a new one—big time. I still don't know why he freaked like that. My head hurt so bad at the time I don't remember much of what went on."

"Your boss here is going to buy us lunch." Boniface had moseyed over to listen to their conversation. "I think what I have to say at the restaurant might shed some light on Conrad's behavior. See y'all at Alice's in fifteen."

Denny smiled and nodded. Greg and Boniface took off in short order, the younger of the two driven by an insatiable need for calories. Before leaving to join them, Denny approached Sheriff Wade's car. The officer was chatting with the two young people whom he'd had the pleasure of driving to and from Santa Fe. Denny spent a few minutes talking with them before heading off to rejoin Boniface and Greg.

XVIII

Alice's was quiet. Gone were the crowds of news people who'd descended upon the diner the month before. Today the three surveyors had their choice of tables, and they selected one against a window that yielded an impressive view of the high, snow-capped peaks beyond Taos. Jenny wasn't on duty, but their orders for coffee, and a Coke for Greg, were taken by a competent replacement: an affable and heavily-pierced young woman with spiked, orange-tinted hair.

Boniface stared after the waitress as though she were some sort of freak zoological specimen before clearing his throat to speak.

"Greg, you were wondering what could have made Conrad behave the way he did. Well, the short answer is, of course: money. He was after something that was conceivably worth a fortune, and the only obstacle that stood in his way was an unknowing and unwitting Maria Benavides. That document Maria produced for us, the one she'd hidden in her devotional statue, was the most valuable, exciting thing I've seen in my sixty years as a surveyor—something capable of driving a man like Conrad to murder. In order to comprehend its significance, though, one first has to understand and appreciate the political and cultural histories of the great State of New Mexico. Every boundary surveyor worth his salt should know about this. At the risk of being pedantic, I feel it's important to set the stage for what I want to reveal."

Their server returned to take their food orders. Denny and Boniface ordered light fare. A nutritionally deprived Greg decided to pull out all the stops and went with the large 'Taco Grande' combination plate. Guaranteed to satisfy the largest man-sized appetite, it was the most expensive item on the menu.

The waitress smiled as Greg placed his order. “Okay, let me shee if I got this shtraight,” she said. In the interest of thoroughness, the young woman wanted to confirm their orders. The task was made difficult by a self-inflicted speech impediment—a large ball-bearing on a stud, skewered through the center of her tongue. “Thatsch one schicken enshillada, one bowl of green shili posholee and a taco plate.” Nods of approval prompted a metallic smile as she took their order to the kitchen.

“That’s disgusting,” Boniface scowled.

“Oh, I don’t know,” Greg said, “If you can look past the trimmings, she’s kinda cute.”

“Humpf, I don’t see ‘cute’ at all. What I see is a walking advertisement for a hardware store, or maybe a poster child for how to tempt infection.” He shook his head in wonder, “The older I get, the stranger everyone around me seems to be.

“Well, getting back to business,” he continued, “the history of New Mexico has always fascinated me. Contrary to what most Easterners believe, France and England were not the first European powers to force their way into North America. Beginning with Columbus, the Spanish launched numerous expeditions to South and Central America, and by the mid-16th century, Spanish troops, Los Conquistadores, had pushed northward all the way through New Mexico and beyond. All Europeans, of course, had one practice in common. They laid claim to any territories upon which their representatives set foot—the rights of indigenous people notwithstanding.

“As a result of these explorations, regions of settlement throughout the American Southwest were established. Relations between the invading Europeans and the natives were at times cordial, but there were also periods of violent conflict. One notable hostile interaction occurred in 1680 and is known as the ‘pueblo revolt.’ Tired of efforts by the Spanish, especially their Catholic Priests, to alter their cultural ways, numerous Indian Pueblos allied to expel the Spanish from their lands. The revolt succeeded. Spanish settlers were driven from northern New Mexico for a period of about thirteen years. Inexorably, however, the settlers returned, accompanied by Spanish troops, and for almost a century and a half thereafter, Spanish and native Indians went about their ways, gradually mixing or, at least, coexisting.

"In 1821, settlers in this part of the continent had developed enough of an identity that they seceded from Spain and established the nation of México. Now, to say that things didn't go well for México in her northern territories is a gross understatement. She didn't have the population or the military strength to hold onto her claims. The Texas fight for independence in 1836, the US-Mexican War of the 1840s, and the Gadsden Purchase of 1852 transferred almost all of what are now California, Arizona, New Mexico, Texas, Colorado and Utah to US control. The term 'Manifest Destiny' is little more than a euphemism for what some call the largest hostile land grab in US history." Boniface paused to sip his coffee and reflect for a moment. He shook his head, "Might makes Right," he said sadly. "Unfortunately, that's how land has been acquired throughout history. And all of this leads up to a very important document."

"The one in Maria's possession?" offered Greg.

"No," Boniface smiled, "'The Treaty of Peace, Friendship, Limits and Settlement' between the United States and México: the formal treaty that ended the Mexican War. It's more commonly known as the 'Treaty of Guadalupe Hidalgo,' or as I call it: 'GH.' Signed in 1848, it contains provisions which continue to cause problems to this day. You see, the US government agreed to recognize and uphold the property rights of the citizens who lived within the conquered territory. Difficulties immediately resulted because of the way the US on one hand, and Spain and México on the other, regarded land 'ownership.' In our country, land is viewed as a commodity, and proof of ownership is documented by a deed and a survey description. By contrast, in old Spain and even in old México, land was something occupied by the citizenry to be developed and to create prosperity. Nonetheless, many documents, which are now collectively known as 'land grants,' were often issued to individuals and groups by both Spanish and Mexican authorities. These land use documents typically contained references to natural landmarks and often gave bearings and distances to define extent. Thus, they constituted surveys. To the American way of thinking these were essentially 'deeds,' and they were given considerable weight in decisions dealing with ownership.

"The US government gave two successive entities the authority to review and approve land ownership claims. The first of these was the Office of the Surveyor General of the Territory of New Mexico created by Congress in 1854. From 1891 until 1904 that adjudication authority was transferred to the newly created 'Court of Public Land Claims,' aka the CPLC.

"Because of the more relaxed attitude of the Mexican government towards land ownership, the claims people brought to the Surveyor General's office often carried with them little documentation, so occupation or habitation of the land weighed heavily in one's favor, especially in and around villages. You see, although some land grants were assigned to individuals, many of them were communal, originally intended to inspire the establishment of towns or 'centers for commerce and defense.' As a result, numerous families might benefit from a single grant if they could prove long-term habitation on the land. Approval rates were high prior to the Civil War, but afterwards they declined. Increased scrutiny crept into the process in order to combat cases of corruption and fraud."

"I read about something called the Peralta Grant," Greg interrupted. "Some guy from Saint Louis fabricated a royal-sounding name for his Hispanic wife and then, by altering records and forging documents, tried to lay claim to twelve million acres."

Boniface smiled, "Yeah that was pretty audacious. Sadly that case and others made the Surveyors General and the CLPC paranoid in their work. By the way, the fellow you refer to was named James Reavis, and he earned himself a couple years in the slammer for his efforts. But, Greg, many of the supposedly fraudulent claims really weren't. They were legitimate petitions based upon subjective interpretation of scant, generally unwritten, evidence. Given the stringent evaluation criteria employed in the 1870s and beyond, there's no doubt that many citizens were denied large acreages to which they were rightfully entitled. And what fraud existed wasn't limited to the common man. Henry Atkinson, Surveyor General of New Mexico in the late 1870s, tried to assign the nearly 400,000 acre Anton Chico grant to a land and cattle company of which he was a principal owner. He and his cohorts presented a convoluted argument involving an original Mexican grant that was later modified by a litany of sales, transfers, foreclosures and overlapping claims in an attempt to wrest the land from its rightful owners. He lost in court when many of his supporting documents were shown to be 'highly questionable.'"

Lunch arrived. The three hungry men dove into their orders along with sides of corn tortilla chips and home-made salsa. "Oh, God, I've died and gone to heaven," Boniface sighed after relishing his first spoonful of posole. "That chili sure hits the spot: meaty, flavorful and just hot enough to bring tears to my eyes and a song to my heart."

They ate in silence for several minutes before Denny spoke. "I appreciate the history lesson, joven (youngster), but what does this have to do with Maria's document?"

His mentor tore off a hunk of freshly prepared sopaipilla and used it to mop up the remaining sauce from his soup bowl. "It was an old land grant, Niño." He popped the sopaipilla morsel into his mouth and washed it down with black coffee. "It's the most historically important land conveyance it has ever been my privilege to behold." He paused in remembrance of what he'd witnessed. "Right now it's in the hands of a respected and trustworthy friend of mine in the Archaeology and Antiquities Department of the University of New Mexico. He's running tests on it to determine its authenticity, and from what he told me yesterday, it's the real deal."

"So why's it so important?" an impatient Greg asked.

It's an old, and I mean REALLY old Spanish grant. It dates from 1671." He paused, awaiting response from his colleagues.

"That precedes the pueblo uprising of 1680," Denny noted quietly. "I thought there were no records that survived the revolt."

"Not many," Boniface agreed. "The grantee, a fellow named Javier Bustamonte, must have taken it with him when he was forced to leave. But he returned to this area when hostilities ceased, and his descendants have lived here ever since. There were several accompanying documents that Maria saved along with the grant, documents written much later—probably in the 19th century—which describe the ranch property in more detail, giving ancillary information such as the number of head of sheep, and later cattle, that were raised and sold, etc. They were undoubtedly prepared as part of a submission package to either the Surveyor General's office or, more likely, the CPLC."

"So, you think the grant was presented as evidence and the claim was rejected?" Denny asked.

"Yeah, I'm pretty sure. What I find to be almost as important as the land grant is the Benavides' family Bible. As you might imagine, that's something Maria was reluctant to part with, but she let me make photocopies at the nursing home of pages in the back that contain carefully scribed genealogical data. If a claim was submitted it would have been in the late 19th century, probably by Maria's grandfather, Julio Peralta. Peralta was her maiden name, before she married Augusto. There are scholars who can find out if the Peralta clan submitted a claim, and if they did, why it wasn't approved. It might have somehow been mistakenly associated with the fraudulent Peralta claim that Greg mentioned. Who knows? We may never learn the truth. I'm almost certain the CPLC was involved at some point because when Rosa and I first spoke with Maria, she said that the 'ladrones,' or thieves, who worked for the government tried to take their land and were fired from their jobs. I think she was a bit confused. The CPLC was disbanded in 1904 and her grandfather ended up having to buy their ranch, or what was left of it, from the Government Land Office one year later. In a way, though, she's right; denying a legitimate claim is tantamount to theft. All I know is, those stuffed-shirt CPLC justices didn't approve many requests. They had a difficult job, there's no doubt about it, but I feel a lot of hard-working farmers and ranchers got screwed as a result of their indecisiveness."

The bitterness in Boniface's voice changed abruptly to excitement as he switched topics. "Now, compadres, let me tell you what my buddy, Steve, the archeologist at UNM, found out. He damn near shit when he saw the grant. He said that the verbiage and word usage are vintage Old Castilian Spanish. If the document were a forgery, it would presumably have been produced after 1854, and there would have been few, if any, crooks capable of writing with such linguistic nuance. Also, the Spanish official who issued the award, the local Alcalde Mayor, was a man named Domingo Bazan de Colon. Steve found a digital copy of Bazan's signature on file at the University and guess what? It's a perfect match to what's on Maria's grant."

Boniface signaled their waitress for more coffee and turned his excitement level up a notch. "And it gets better: Steve has a special laboratory where he and his colleagues prepare samples for carbon dating. Apparently, one has to be very careful not to contaminate them with modern carbon sources: things like sweat or skin cells from one's fingers, bits of dust, etc. They have to handle specimens in an ultra-clean facility. Anyway, they sent samples to a lab in Canada that does something called accelerator mass spectrometry. Steve put a rush on the job and got the results back a few days ago. Both the parchment and the ink that was used are at least three hundred years old. And," he paused for effect, "the chemical constituents of the ink are appropriate for the period. Gentlemen, that land grant is authentic!"

Greg pushed his empty plate aside. "Okay, so you've gone to a lot of trouble to prove the grant is legitimate, but why? I mean, aside from its historical significance, what does it say?"

"It says that Maria Benavides and Conrad Sussman, as the sole living heirs of Javier Bustamonte, own almost 15,000 acres of local real estate. The language in the grant is surprisingly specific and clear. It bequeaths to Bustamonte all that lies within five leagues of the paraje San Cristobal, except to the north and west where the grant extends across the Arroyo Castaneda to the foot of the red hills beyond. What I feel lends further credibility to the grant is that it recognizes the boundary of the Santa Teresa reserve. The Spanish were 'kind' enough to issue grants to many of the pueblo Indians whom they were trying to subjugate and convert to

Catholicism. The Bustamonte conveyance was probably signed at the same time as a grant to the adjoining pueblo—a document which undoubtedly meant nothing, at the time, to the Indians, and almost certainly no longer exists. I imagine the pueblo boundaries were probably re-established by parole testimony after the 'GH' treaty was signed."

"Whew, so if the CPLC screwed up and didn't approve the grant—which you've proven to be legitimate—what happens now? Do Maria and her dipshit nephew own 15,000 acres of land, including the entire town of Arenas Rojas? Hell, that means they own the restaurant we're in right now."

Denny weighed in, "That's right, Greg. In an unusual case like this, it's up to the courts to decide what to do. It's likely that Maria will be reimbursed by a combination of monetary payment for lands that have been sold to others and the transfer of any federal lands which lie within the grant. I don't know what Conrad will get if he's a convicted felon, but the motive behind his recent actions is readily apparent. He's probably known about the land grant for some time. The frequent visits to his aunt and uncle's ranch over the years certainly could have made him aware of its existence. His claims of complete ignorance of the Spanish language are also, I believe, exaggerated. When he volunteered to handle the land sale to the pueblo, it put him into a position where he could monitor what documents Augusto presented to the court in protest of the transaction."

"But, he represented *Dirk* in that sale," Greg responded, "that makes no sense. You'd think if he knew about the grant and its provisions he'd have been a champion for the rights of Augusto and Maria, his blood relatives."

Denny's mind was working fast, the facts were tumbling into place. "No, Greg, he didn't care at all about the sale to the pueblo. He wanted the grant for himself. Representing Dirk was a clever way to force Augusto into presenting the grant in court—in a desperate attempt to prove ownership. Once the land grant surfaced, Conrad would immediately have thrown his support the other way, behind the rights of him and his family. Proceeds from the grant would dwarf the million or so dollars resulting from the Santa Teresa Pueblo deal. And I've no doubt that Augusto and Maria would have met with a fatal 'accident' sometime after the grant was made public, thus transferring the assets to Conrad."

"Augusto was too smart for his bastard nephew," Boniface interjected. "He suspected what Conrad was after and he made Maria promise never, under any circumstances, to tell him where the grant was. Even after Augusto's death she refused to give Conrad the information. She confided to Rosa that her promise to her husband was still in effect. Conrad was never to see 'the papers.'" His temper flared. "Augusto knew what an asshole Conrad is. He saved Maria's life by denying that prick what he wanted. But I have to ask the two of you to never tell Maria that. She has trouble enough comprehending that her sweet 'sobrino' is in jail."

"I'm really confused," Greg admitted. "First, Bella does her thing. She moves a corner monument, alters a deed, kills Augusto and frames my boss for the murder. Then Conrad enters the picture, representing Dirk, not his own family, in the associated land sale. He kills Emilio, who has nothing to do with any of this, ransacks the Benavides house, kidnaps me, Lily and Xeno, and now this land grant pops into the mix. There's so much that went on—and neither of you suspect that Bella and Conrad were working together? Come on, I just don't get it."

Denny laughed. "Everything's really very simple, Greg. You, yourself, just mentioned the key: there were two separate crimes here, totally unrelated to one another. Once you accept that fact, everything falls into place.

"Bella was strictly concerned with fleecing the Benavides family out of land that was worth a fortune to the pueblo. She rigged it so that Dirk would sell the property—something that, if he really owned it, would have been his asset per their pre-nuptial agreement—and deposit the funds into a joint bank account. She would then transfer most or all of it into one of her private accounts before she and Dirk inevitably divorced. Believe me; Dirk would never have caught on. Conrad effectively played no role in any of this except for Bella's recommendation that he serve as my advocate. She knew he was no criminal defense lawyer, and she knew of his cocaine habit. With him defending me against the charge of killing Augusto it increased the odds of me taking the rap.

"In contrast, Conrad's goal was to flush the land grant into the open. He didn't care at all about the land sale to the pueblo, the status of Dirk's deed, or anything else that Bella was doing. When Augusto was killed, however, he became extremely nervous. As far as he could tell, Augusto had been the guardian of the land grant. Maria either didn't know where it was or wouldn't tell him. He suddenly became suspicious of anyone who'd been in contact with her. Remember our phone conversation right after I was released from jail? You told me that you and Emilio had just spoken with Maria about the mysterious papers she'd referred to. You also mentioned that Conrad was present at that interview. Well, I believe that's when he freaked out. Right after the meeting he went to the stables to confront Emilio, demanding the whereabouts of the grant. The ranch hand, of course, had no idea what Conrad was talking about. He probably dismissed him with a suitable off-color response, but when he turned his back, Conrad killed him. Our favorite lawyer then proceeded to ransack the Benavides property, and then, not having found what he was after, he tortured you, Xeno and Lily on the chance that one of you might know something.

"It really saddens me that there was no way to foresee any of this. Augusto and Emilio were fine people. I hope the legal system gives Bella and Conrad what they deserve, shows them no leniency."

Lunch was over. Both Boniface and Greg were anxious to get home. They said their goodbyes and left immediately. Denny harbored the same sentiment, but there was one more task he had to perform before he could leave. He ordered more coffee and pulled out his cell phone. Following a brief conversation, he sat back down to wait for Bobby Wade to make the drive from Taos.

"THANKS AGAIN FOR doing this, Sheriff," Denny said.

"Hell, this ain't no favor. It's part of my job."

"Did you get the warrant?"

"Yep, the Judge's signature ain't even dry yet. You lead the way and I'll follow."

Denny once again found himself outside the massive gate to Dirk's ranch. He and the Sheriff drove amidst snow flurries to the parking area in front of Dirk's house. Denny hustled to the stable area to corral Lily and Xeno, and the three of them joined the Sheriff as he knocked on the door. To reinforce the official nature of his visit, Sheriff Wade had left his patrol car running with its lights flashing. Both Dirk and Larry answered the door.

"Sheriff, what's the meaning of this?" Larry asked irately, "Such an unannounced visit is most inappropriate . . . ."

"Lawrence Cooper, I have a warrant here for your arrest."

"My what . . . what the hell are you talking about?"

"It says you're being charged with embezzlement, sir."

"Well, obviously there's been some mistake." He looked to Dirk for support. "Mr. Stetson here can certainly vouch for my honesty, sheriff. Either someone has falsely accused me, or there has been a terrible misunderstanding." He levied an icy stare at Lily. "I suspect I know who's behind this."

"That may very well be true, Mr. Cooper, but it's something for a judge to decide. It's my sworn duty as Sheriff of Taos County to serve this warrant and to arrest you on suspicion of felony grand larceny."

Larry stomped his foot. "I'm not going with you, sheriff. This is an attempt to arrest an innocent man under false pretenses." He turned once again to address Dirk, whose countenance had become uncomfortable and pale. Denny thought the man was going to pass out. "Please, Dirk, explain to this jackbooted thug . . . ." When Larry looked away, Bobby Wade whipped a set of handcuffs from his belt and had Larry's hands pinned behind his back within seconds. Larry howled in rage and launched into an obscene tirade. However, he suddenly winced and went quiet as the Sheriff applied a slight twist to one wrist.

"You have the right to remain silent, Mr. Cooper, and I strongly suggest that you exercise that right."

When the Sheriff and his charge were gone, it was Lily who broke the awkward silence. "Dirk, Larry's been stealing from you for quite some time. About a year ago I ran across a false invoice he'd prepared for saddles and tack—items that were never purchased. Emilio and I subsequently searched for and uncovered a number of such receipts, and we contacted the vendors who supposedly had made the sales. None of them had any knowledge of those transactions. I know we should have said something sooner, but we weren't sure about what was going on until . . . until right before Emilio was murdered. Earlier today Mr. Carswell suggested I take the evidence to the Sheriff." She bit her lip and fell silent.

Denny said, "Mr. Stetson, I know you're a busy man and need help running this ranch. Furthermore, I realize what a shock it must be to find out that an employee, whom you'd grown to trust and depend upon, was stealing from you. But there's another of your employees, standing beside me, who is more than up to the task of replacing Larry." Lily blushed at becoming the center of attention. "You may not be aware of this, Dirk, but Lily is quite a scholar. Her mathematical skills are extraordinary, and her command of the English language is the best I've ever seen—she's a walking thesaurus!"

Lily beamed a smile that would warm the coldest heart. Dirk responded accordingly. "Well," the actor said decisively, "I don't know about the dinosaur part, but she sure knows horses, that's for darn sure. Lily, I'd really like it if you and, uh, this other young feller would stay on and run this here ranch. Wha'dya say?"

"I really appreciate your estimable proposition, sir, however, Xeno and I have accepted alternative labors."

After a brief silence, Denny said, by way of translation, "You've got another job?"

Xeno responded with enthusiasm. "Yep. Old Lady Benavides beat you to it, Dirkaroony. That other surveyor dude, El Pedo Viejo, suggested we help out around her ranch until she gets back on her feet." He put his arm around Lily and hugged her. She blushed deeply but made no effort to pull away. It was apparent to Denny that their relationship had entered a new phase. The show of affection warmed his heart. Deep down, Denny was a hopeless romantic, and he felt privileged to stand in the presence of two young people who he knew were destined to be together. “This wonderful gal wants to help that nice old woman, and I plan to stay right by her side. She means more to me than, well . . . let’s just say that Lily is to my soul what Euler’s formula is to mathematics. She’s a unifying entity that is beauty, poetry—absolute perfection!”

The statement brought a tear to Lily’s eye. “That’s the nicest complement, (sob) . . . Wow, Euler’s Formula!” Quickly regaining her composure she addressed Dirk, “I’m sorry we have to leave your employ, Mr. Stetson, however, I’m sure we can negotiate an agreement if you wish me to continue to manage your equestrian facilities.”

Denny again produced a tactful translation, “Well, how about it Dirk? Can she still take care of your horses?”

XIX: The Following Spring, March, 25th

"Hurry up, Denny, or we'll be late."

"Relax, it's only a forty-five minute drive to the theater and we've got over an hour. Besides, there's going to be a lot of ceremony prior to the movie. We can miss most of that."

"You're kidding, right?" came the incredulous reply from his wife, Pamela. "This is a World Premiere and we're distinguished invited guests. It's a once-in-a-lifetime experience."

"It's just a movie."

"Oh, come help me with this dress, will you?" Denny put down his newspaper and padded in stocking feet to where his spouse stood in front of a full-length mirror. After helping her shimmy into the rather tight-fitting garment, his hands drifted to familiar places. "Not now, you randy, old goat," came the stern but loving rebuke, "there'll be plenty of time for that after the movie. You'd best get dressed. Oh, and correct me if I'm wrong: Bella was sleeping with the producer, what's his name?"

"His name's Andre, and he's the director."

"And Dirk was sleeping with his co-star?"

"No, her name's Lavern and she was a script reader or some such thing. The gossip is that she may soon become the fifth Mrs. Stetson."

"Really?!"

"Yep."

"And Dirk's real name isn't Stetson, it's what?"

"Hertzfelter, Woody Hertzfelter, and Bella's real name, before she legally changed it, was Agatha Cruz."

"Well, I can certainly see why they changed their names."

"Why?"

"Oh, Denny, you have no sense of style. Hertzman . . . ."

"Hertzfelter."

"Right. That's not a name for a ruggedly handsome leading man. And Agatha sounds more like someone's spinster aunt, or a mystery writer, not a big screen starlet."

"You know, babe, there were more important things going on at that ranch last fall than people sleeping around—murder, theft and corruption, to name a few. And who gives a shit about a person's name as long as he can act?"

"I understand, sweetie, but all that's over and done with, and movie star gossip never ends. All I can say is that it's good you became a surveyor and not a Hollywood publicist or we'd be panhandling for a living. Can you zip me up, please . . . and watch those roving hands."

"There you go. You look fabulous, my dear."

"Thank you." She gave him a kiss. "Now, get dressed."

"I am dressed."

She gave him a deprecating look, "You *can't* be serious."

"It's going to be dark, and people are going to be watching the movie, not us. And anyway, this shirt is brand new. Greg gave it to me last week for my birthday."

"It says, 'Aggie Basketball'!"

"Yes, and it has a collar and everything." Denny's tone conveyed defeat. He knew the argument was over. Relocating to his dresser he began rummaging, looking for a shirt that would pass muster.

DAYLIGHT PICTURES SET a general release date for 'Conestoga Highway' of April 1, 2013. The premiere showing would be held at the Lensic Theater in Santa Fe one week before then. Santa Fe had been selected for the first public viewing because the film had been shot entirely in New Mexico, and the majority of the scenes had been set in and around the capital city.

When he and Pam arrived for the premiere, Denny had to admit that his wife was right. "You'd think we were going to the coronation of a king," he said in amazement when he saw the quantity of fine furs and jewelry adorning the patrons who waited patiently for the doors to open. Once inside, they felt sorely out of place. Almost everyone else seemed to know one another, and the Carswells kept to themselves until Denny spotted Sheriff Wade and his wife. The two couples made small talk for several minutes before the Wades decided to take their seats. Denny had promised to introduce his wife to Dirk, who had yet to arrive at the theater, so they opted to wait in the lobby for the star's grand entrance. Moments before the program was slated to begin, a long, black limousine pulled up out front and disgorged members of the cast and production crew who had made the trip to Santa Fe for the premiere.

In the interval from November, 2012 to the following March, Dirk Stetson's status in Hollywood had skyrocketed. He had become, without doubt, filmdom's leading man. Ironically, Bella's arrest had not harmed his career in the least. Instead, clever publicists had used it to garner public sympathy for Dirk. She was vilified in the tabloids as the dastardly, cheating wife, plotting murder and theft while her loyal, unsuspecting husband worked hard on the set of 'Conestoga highway.' Insiders who knew better went along with the story because, in general, the sentiment was: 'what the heck, she is a murderer, for heaven's sake.'

Dirk was further elevated towards sainthood by the heroic wound he'd suffered while wresting a gun from a killer who'd gone on a shooting rampage on his neighbor's ranch. The cut on Dirk's face hadn't required stitches, but an astute publicist had insisted he get several of them. A local physician in Arenas Rojas was handsomely reimbursed to add half a dozen prominent stitches to the 'wound' using highly-visible, jet-black silk. Numerous photos were taken. When filming resumed after the incident, make-up artists were instructed to allow the sutures to be barely visible in close up shots of Dirk's face. Via a well-orchestrated leak from a 'reliable source,' the scandal magazine 'Hollywood In The Know!' paid an exorbitant sum for 'clandestine' photos taken on set of Dirk's cheek. A wonderful companion article informed the public to look closely for the hero's scar towards the end of the film. Executives were salivating as a result of the serendipitous events that occurred at the Stetson ranch during filming. To capitalize on those events, it was decided to release the film much earlier than originally planned. It was no coincidence that the premiere was showing on the same day as jury selection for Bella's trial. Ticket sales were expected to soar into the stratosphere.

"There he is," Denny said in a hushed voice. "C'mon, I'll introduce you.

"Hey, Dirk, it's good to see you again. Your cheek seems to have healed nicely."

"Yes, sir, they put me back together good as new. Why it's good to see you again, Mr. uh . . . ."

"Carswell, Denny Carswell. I'm the surveyor who worked on that land sale last fall. I'm sorry about the mix-up with your deed, and, well, with everything else that happened. By the way, Dirk, this is my wife, Pamela."

"Why, it's a pleasure to meet you, ma'am, a real pleasure. I want you to know that without your husband's skills on our set, 'Conestoga Highway' would never have become a success, no siree."

Pam couldn't think of what to say. Fortunately at that moment the theater lights dimmed, and Dirk let her off the hook. "Well, it looks like they're startin' the show. I gotta go; Andre's gonna interview me 'fore the film starts. Hey, it was a pleasure to meet you, uh . . . ."

"Pam"

"Yes, Pam, and Mr. uh . . . ."

Denny heaved a sigh. He was mildly offended that Dirk still didn't seem to have a clue who he was. "Dennis," he replied, "Dennis Rodman, from the NBA, you know, the 'National Boundary Agency.' It was an honor for me to have helped such a major film star with a boundary problem."

"The pleasure is all mine, Mr. Rodman. It's good to see you again."

Dirk broke away and began to walk toward the stage, smiling and shaking the hands of adoring fans as he went.

"That was rude of you," Pam admonished.

Denny nodded, "Yeah, I guess I just got fed up with his inability to remember me. After all that happened last fall—hell, he's forgotten what I said already. Let's go get a cup of java, we've got a few minutes still." He and Pam went to the concession area. Over the PA system they could hear the dialog between Andre Bivens and Dirk, the interview that was to be the lead-in to the film.

". . . and after all that happened you were still able to concentrate on your job, to focus on your lines. That's amazing, Dirk, truly amazing, the mark of a real professional."

"Shucks, Andre, you know what they say in our business, 'the show must go on.'"

"Indeed, Dirk, indeed. Now, tell us . . . ."

For the most part, Denny tuned out what was being said. This stuff was just fluff, fodder for public consumption. The lobby was nearly empty now; most audience members had taken their seats, and he and Pam were among the few stragglers left in the lobby.

Denny took in his surroundings. He loved the Lensic Theater. It was a throwback to the gilded age of the twenties, with ornate trim, plush carpeting and balconies, like the one they would soon occupy, that brought the viewer close to the actors in live performances. Although he himself wasn't much of a movie buff, Pamela was, and he vowed to bring her here more often. The interview appeared to be winding down, so he and Pam began to maneuver towards the stairs and to the balcony reserved for special, invited guests.

"Well, Dirk, it's clear that the strange and tragic events that transpired during the making of this film are unprecedented. Do you have anything more you'd like to add?" Andre asked.

"I sure do. Ladies and gentlemen, I'd just like to thank everyone who helped make this film possible."

"That's very thoughtful of you, Dirk. Now, without further delay, let's . . . ."

"What I mean, Andre, is the little people who worked on the movie. I don't mean people who are small, no sir, I mean people who aren't actors, like the cameramen, sound men, directors; people like you, Andre . . . ."

"That's very kind of you, Dirk, but let's move on now and show our audience the final product of last year's efforts."

". . . and Wendy," Dirk continued, "and the folks who helped with the sets. Many of them are here tonight."

"Yes, they are, Dirk, but . . . ."

". . . and I'd like to thank my good friend Dennis Rodman of the NBA, who . . . ."

Denny spluttered and nearly spilled his coffee. Had he heard right? Did Dirk just say what he thought he'd said? Confirmation came from his wife. "Gee, Mr. Rodman, that sharp-witted actor remembered your name after all. Congratulations."

Her sardonic attitude changed to concern when she looked at him. His face was buried in his hands and he was shaking violently. "Denny," she asked apprehensively, "Denny, are you all right?" A nod and a muffled guffaw escaped as he finally drew a breath. He was laughing hysterically. "I don't see why you're laughing. When the press pick up on this they'll . . . ." Denny was trying mightily to regain his composure, but he just couldn't stop laughing. The one time Dirk remembered—the one name he'd been able to carry on his tongue for more than a heartbeat—it was the strangest, most absurd, and quite possibly the most wonderful thing he'd ever heard—simply freakin' hilarious.

Pam was soon caught up in her husband's mirth. After all, there was nothing she could do to change what Dirk had said; it really was ludicrous. She chuckled and said, "My, you've certainly had a busy few months, Mr. Rodman: first a diplomatic junket to North Korea to explain the finer points of rebounding to a pudgy despot, and then a visit to restore the memory of a dimwitted movie star. That must be very satisfying."

Denny was in tears, now. He hugged his wife, "Yes," he stammered, "Yes it is."

"All kidding aside, though," Pam continued with mock sincerity, "I see only one way out of this predicament."

"What? What's . . . that?"

"As I see it, you're going to have to change your name." Denny slapped his thigh and roared with laughter. "Pamela

Rodman," she continued, now laughing heartily herself, "It has a nice ring to it. But do you know what's reeeally funny?" Denny shook his head. "Rodman—get it? What more appropriate surname can there be for a once-highly-respected-turned-smartass surveyor?"

XX: Epilog

The sky was so overcast it seemed that nightfall was imminent, even though it was still early afternoon on a somber Sunday in April. A light mix of snow and hail came and went as moisture-laden clouds swept over the hillside. Lily had come prepared. She held an umbrella over herself, Maria Benavides and Rosa Romero. The two men braved the elements, pulling their collars in tight to keep moisture off their necks.

"That's where I found it," Xeno pointed to a mound of loose rocks a dozen feet from the survey marker atop Cerro Medio. "I was moving that fence post down to where you set the boundary to the casino property. All of a sudden I saw this black thing and then I saw bones and I thought, like, maybe I should call the cops, but when I saw how old this is . . . that's when I figured I'd call you, Surveyor Dude."

Bonifacio Romero knelt to examine the find: an old wooden box, falling apart, with black metal hinges and clasp. Closer inspection showed the hardware to be made of silver, heavily tarnished and elaborately engraved. He carefully pried it open and pulled out a small gold, jewel-encrusted crucifix. The object seemed strangely out of place. Its smooth golden surface gleamed like the day it was made, and the emeralds at the ends of the cross piece were a brilliant green despite the subdued light of this miserable day. Boniface held the piece up close. "I'll be damned," he muttered, and handed it to Maria.

The matriarch of the Benavides ranch smiled and accepted the item. "Gracias." She turned it over with gnarled fingers and held it out for Rosa to see.

"It bears the initials 'IBA' on the back," Boniface said. "It has a date as well, something I couldn't make out in this light." He looked at his wife and saw recognition come over her.

"Why, Boniface," she exclaimed, "it can't be! Maria, this belonged to Javier's wife!"

Boniface nodded. "In your family Bible, Maria, it lists your ancestor, Javier Bustamonte, and his wife, Inez Bustamonte de Anaya. She died in 1680 and was buried at a place called San Cristobal. This is her grave, Maria. I'm certain of it."

The old woman stood holding the cross. Although she clutched it firmly, she no longer regarded it. Her gaze was distant; she seemed lost in thought. The wind stirred wisps of grey hair along her temples. Her angular features, set in profile against the slate-grey sky, were defined by generations of mixed Spanish and Native Indian descent. She stood firm against the cold, this old woman, the last of a proud lineage of hardy souls who'd toiled hard to forge their lives in the harsh climate of this remote corner of New Mexico. Her dark, sunken eyes were bright and alive as she contemplated the meaning of the object in her hands and the incredible history behind it.

The others in attendance understood the significance of the moment. Only the whistling wind dared to disturb Maria's reverie. Finally, she spoke. "Esta bien. Toma lo, por favor. Please take it, Boniface. Put it back. It belongs to her; it was her wedding gift."

Boniface carefully replaced the crucifix in its box and set it gently back into the rock pile. He motioned for Xeno to help him cover it well with dirt and rocks. Then, as the wind picked up and the snowfall intensified, the small party slowly made its way down the slope of the Cerro Medio.

FINIS

Made in the USA
Monee, IL
04 November 2026

67379297R00138